La Boulangerie
Baking at home
with Grégoire Michaud

古風歐陸麵包

Grégoire Michaud

萬里機構・飲食天地出版社出版

Foreword from Celene

I believe a good pastry chef is one who is disciplined with the exact measurements of the ingredients and understands the chemistry that occurred between the yeast, wheat and sugar etc with the aim to create the best product.

An OUTSTANDING pastry chef, however, is one who has mastered all the skills of a good chef and more - he has PASSION. Gregoire Michaud is definitely not lacking in passion, which sometimes borderlines madness. That is why his creation is always interesting, and often courageous with a hint of edginess.

Yes! Beneath the polite sensibilities of this Swiss chef, there is a rebel... with a cause.

I had the privilege of getting to know Gregoire through his contribution in Giving Bread, a non-profit organization which collects fresh and leftover breads from restaurants, hotels and bakeries and give them to the needy, especially the neglected elderly.

Gregoire is one of Giving Bread's earliest donors. At least once a week, I visited him in his kitchen to collect the gourmet breads which he donates to Giving Bread - a delicious spread of croissants, muffins, danish, hearty breads etc. He would plan ahead of time whenever he knew I was coming to collect breads and often accommodate my requests for more supplies when we received special needs from elderly centers.

Once he asked to volunteer in our distribution activity as he was not contented with only baking and giving us the breads. He wanted to go to the slums. On that particular Sunday afternoon, I took Gregoire and a group of volunteers to the infamous cage homes where old men lived in conditions much worse than domestic dogs. We also went to a homeless shelter and an elderly housing estate. When we had finished distributing all the packs of breads, I noticed that Gregoire's ears were crimson and his eyes teary. He was visibly moved by the plight of the needy and sped home to write a heartfelt article about Giving Bread in his blog.

Yes! Gregoire has PASSION. And that's the secret ingredient in his breads which make them so heavenly.

Celene P. Loo
Founder, Giving Bread
www.givingbread.org

序言

我深信一個好的糕餅師，應懂得精準計算材料，並確切明白酵母、小麥和糖融合後的化學作用，才能創造美味的糕點。

一個傑出的糕餅師，除了精通良好糕餅師應有的技術外，還要比一般餅師懂得更多。閔言樂不但熱誠滿腔，有時其熱愛程度還接近瘋狂。他的創作來自其興趣。勇於嘗試的他，產品常常鋒芒畢露，令人眩目。

沒錯！在這個殷勤又觸覺敏銳的瑞士糕點師傅體內滲入的叛逆因子……便是創新意念的動力。

我因服務於非營利機構"愛心麵包"，有幸認識言樂，他代表酒店捐贈麵包予有需要人士，尤其是受忽視的老人。

言樂是敝機構最早捐贈者之一。每星期最少一次，我會到他的廚房收取麵包，品種繁多，有牛角麵包、鬆餅、丹麥麵包、心形麵包等。無論何時，他會預先籌劃，如遇老人有特別需要，他必定體恤地順應他們的要求。

他不但只把麵包捐給有需要的人，還主動參與到貧窮的地方派發麵包。在某個星期日的下午，我帶言樂和一群義工到臭名遠播的籠屋，探訪居住在那裏的老人，那地方的居住條件比狗屋還差，我們也到過露宿者之家和屋村獨居長者之家分發麵包。當工作完成，我注意到言樂的耳朵變紅，熱淚盈眶，為受助者的困境深感難過。言樂隨即在其網誌，與大家分享當天的所見所聞。

對的！言樂的心充滿熱情，這就是他做美妙麵包的秘密材料。

Celene P. Loo
愛心麵包創辦人
www.givingbread.org

Foreword from Xavier

Greg,

It is a real pleasure for me to preface your work. So much more than a book, you are placing an important stone in the edifice of our profession. You are leaving your mark, a trail that others can follow. In writing this book, you are continuing the noble tradition of our forefathers by transmitting their precious savoir-faires.

For you as for me, in France, Switzerland or whatever the country, night time is reserved for our bakers. Our families and friends are sleeping, and discreetly, our work begins. Our master bakers guide us, the kneading machines are running, the ovens are warming, the leavens fill the kitchen with their particular aroma and our commitment to this art starts to rise.

This commitment today becomes part of you, and if you find that as a result of this book, you inspire other bakers, even just through one recipe, by giving them a hint or a simple idea, you have achieved your goal.

I am often in the habit of presenting our profession as the coordination of the hand and the head. Nothing worth doing comes without intense reflection.

The first few pages of the book discuss the theoretic approach of our profession. This is very important. Baking is evolving, and even if it is not necessary to look for a rational explanation for everything, if that aspect of magic subsists, we must do our best to understand, anticipate and plan ahead. Our mistakes must bring us enlightenment and the technology of baking is there to help us.

The values of bread are universal. You continue to show from your home in Hong Kong, that your commitment is intact. The trip you have embarked upon allows your to share you passion with others; it is now up to your readers to discover it.

My compliments to the baker-writer-traveler that you are.

Xavier Honorin
Champion du Monde dela Boulangerie

序言

Greg:

我很榮幸替你寫序。你屬於我烘焙專業的中流砥柱，與這書相比，重要得多了。編寫這書，你把祖先們最寶貴的才幹延續和傳承下去。

你我都一樣，在法國、瑞士或任何國家，午夜時間總是留給烘焙師。當家人和朋友都進入夢鄉，我們卻悄悄地開始工作。在總烘焙師指揮下，各就各位，搓揉麵包機在運作，烤爐在預熱，廚房裏瀰漫著發酵的獨特香味，顯示我們正式投入這門藝術。

時至今天，這份投入已變為你的一部份，你寫出了這本書，你在每個食譜中給予的提示和意見，足以啟發其他烘焙師，相信你已達到目標了。

我習慣以雙手配合腦袋去表達我們的專業；沒有用心的思考和自省，做出來的製品實在一無是處。

這本書前幾頁內容，談論麵包製造的理論，這是非常重要的。烘焙在不斷演變和進化，雖然毋須每個細節都作理性分析，但如果這是魔法的原動力，我們必須盡力理解、估計和預先部署。從錯誤中獲得啟蒙，我們的烘焙技術才會有所進步。

麵包的價值無分地域，你把麵包帶到香港，繼續獻身於這個行業，並與別人分享你的熱情，讓讀者從中感受到製作麵包的樂趣。

我向集烘焙師、作者和旅遊家於一身的你致意！

Xavier Honorin
世界杯麵包大賽冠軍

Preface

Largely drifted from my first baking book Artisan Bread, I have oriented this second edition towards baking your own bread at home with all the advantages, inconvenience and great challenge it induces. I left the complicated techniques and sciences aside for the moment and kept the popular recipes, modified them for easier home baking. I also added other great new recipes – all tested and baked in our very own ovens!

This book has perfect recipes for the novice home bakens; yet, the advanced bakens will also find creative ideas and great quality baking methods to be used in a professional environment.

In this second edition, I thought helpful to talk about the usage of baking stone at home and as well, I have adapted all the recipes using dry yeast instead of fresh yeast, the latter not being the most common in today's grocery stores. The exhaust is also a feature that is rarely found on home ovens, thus I am also explaining that concept. Flour is a typical topic of uncertainty in home baking, and you'll find an easy guide to which supermarket flour is suitable and for what purpose. Additionally, for you, baking experts out there, I have added the baker's percentage in each recipe.

To me, baking at home for family and friends is one of life's most rewarding moments. From scratch, you create an experience, a gift, a memory that will be etched in minds as a fond time. One of bread loaf's greatest features beside of being an amazing food is that it is shared amongst people; bread is convivial. When your bread comes out of the oven and starts its subtle crackling noise, dispersing its fairy tale scents all around home, the magic happens.

Good bread in its classic and traditional form is the ultimate goal of any bread lover; with this, bread is a staple food that crossed centuries without losing any of its appeal – almost untouchable. For that matter, there was never any 'espuma of baguette' or 'rye bread caviar' bred from modern cuisine; and I'm thankful for that.

Today, the values of real bread in our society aren't generally as deeply anchored as in the past; yet, thanks to a solid worldwide movement, we see real bread raising and rising again! I could have written a whole chapter on recent bread development, but I thought Robert Orben summed it all pretty well when he said:

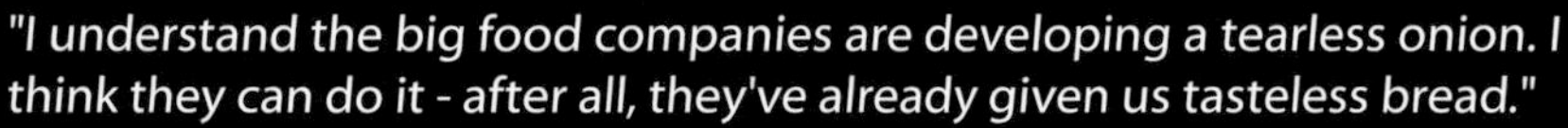

"I understand the big food companies are developing a tearless onion. I think they can do it - after all, they've already given us tasteless bread."

Grégoire Michaud

前言

雖然這書介紹的麵包不少與我前一本作品《麵包教室》相同，但在這書裏，我改以家庭式烘焙為主導概念，撇除不便之處以及可能造成困難的做法；我剔走了繁瑣的技巧和複雜的科學理論，讓這本書成為大眾化的食譜，化繁為簡，方便在家製作。我還加入許多創新食譜 —— 所有的新嘗試皆能從你的烤爐烘出！

這書有很多一流的食譜，特別為新手而設；至於同業，則可從中找到創意，或將優質烘焙方法實踐在專業領域中。

在本書裏，我特別談及在家中如何運用烘焙石以及採用快速乾酵母取代新鮮酵母，因為新鮮酵母不普及，不易在雜貨店找到。除此之外，我又會解釋設有噴射水蒸汽功能的家庭式烤爐。在麵包製作中，麵粉是最重要的成份，為了切合需要，書中提供一個快速指南，以方便大家在超級市場內找到所需的麵粉。我還在每個食譜附加百分比，讓專業烘焙師可自由調校份量。

對於我而言，在家烘焙美食，招待親朋好友，是生命裏的珍貴時刻。從搓揉過程，獲得的體會，是一份禮物，一段深深印在腦海的美好回憶。一條麵包，最偉大之處是能與人分享，亦是一場歡宴不可缺的美食。當麵包從烤爐取出，傳出隱約的碎裂聲響，陣陣的香氣瀰漫在家裏，魔法便開始了。

優秀的麵包揉合了經典和傳統的模式，亦是任何一位麵包愛好者的終極所愛，於是，麵包一直以來都是主要食糧，千百年來都沒有失去吸引力，慶幸還能保持原汁原味，沒有出現 " 分子料理法包 " 或 " 黑麥包魚子醬 "！

今天，真味麵包的價值也許不及從前；感謝在世界各地的積極推動，真味道的麵包終於再度抬頭！其實我能為近代麵包的發展大作文章，不過，借 Robert Orben 以下一句話正好作為總結：

“我明白大型食品公司已研發出不刺激眼淚的洋蔥。我想他們必能做到 ——沒有味道的麵包了。”

閔言樂

CONTENTS 目錄

002 Foreword 序言
006 Preface 前言
010 How does it work? 麵包是怎樣做成的？

A touch of sweetness 甜麵包的接觸

032 Candied Ginger Pan D'oro 蜜餞薑粒皇家麵包
034 King's Bread 皇冠麵包
036 Sweet Potato & Caramelized Chinese Walnut "Cocotte" 地瓜焦糖核桃麵包
038 Pearl Sugar & Chocolate Bread 珍珠糖巧克力麵包
040 Hazelnut Praline Flaky Rolls 榛果巧克力脆卷
042 Pecan and Raisin Boule 葡萄乾胡桃麵包
044 Coconut Palm Sugar Roll 椰糖奶油卷
046 Dark Chocolate and Sour Cherry Bread 黑巧克力酸櫻桃麵包
048 Thick Dried Fruit Loaf 乾果麵包
050 Saffron Vanilla Cuchaulle 番紅花香草麵包
052 Prune Walnut Bread 梅乾胡桃麵包
054 Pear Frangipane Brioche 洋梨杏仁塔

The arrogance of salt 了不起的鹹麵包

058 Portobello Mushroom Country Bread 大蘑菇鄉村麵包
060 Sun Dried Tomatoes & Pancetta Rolls 番茄乾義式培根卷
062 Gruyére Cheese & Paprika Twist 辣椒葛瑞爾起司條
064 Baby Onion and Chervil Bites 法式小洋蔥
066 Tomato Chili Pretzel 德國番茄辣椒椒鹽脆餅
068 Confit Garlic & Oregano Schiacciata 大蒜奧勒岡扁麵包
070 Kalamata Olives Baguette 希臘黑橄欖法國麵包
072 Fresh Rosemary Pavé 新鮮迷迭香麵包
074 Black Pepper & Fennel Taralli 黑胡椒茴香脆條
076 Crispy Pine Nuts Basil Rolls 香脆松子羅勒卷

078 Grenaille Potato and Lemon Thyme Foccacia 小馬鈴薯檸檬百里香披薩
080 Bacon & Herbs Epi 培根香草麥穗
082 Arugula & Parma Ham Pizza 帕馬火腿芝麻菜披薩
084 Russian Pampushka 蒜蓉香草俄羅斯麵包
086 Red Bell Pepper and Bagnes Cheese Rolls 紅甜椒瑞士起司卷
088 Piment d'Espelette Lavosh 法國紅辣椒薄脆
090 Provencal Fougasse 普羅旺斯面具
092 Garlic Parsley Brioche Needle 大蒜洋香菜麵包
094 Cider Raisins, Smoked Bacon and Coriander Rolls 培根葡萄乾小法國

Almost untouchable 跨世紀麵包

098 French Baguette 法式長麵包
100 Buckwheat Pillow Bread 蕎麥枕頭麵包
102 Petite Tresse au Beurre 瑞士辮子麵包
104 Seeded Oval 粿麥雜糧麵包
106 Whole Wheat Irish Soda Bread 全麥愛爾蘭蘇打麵包
108 Classic Ciabatta 義大利拖鞋麵包
110 CRUMB 造型麵包
112 Spelt Farmer Bread 斯佩爾特小麥農夫麵包
114 Light Roselle Rye Boule 洛神花粿麥麵包
116 Caraway Bordelaise Crown 孜然波爾多皇冠麵包
118 Fabrice Country Stick 鄉村麵包棒
120 Horseradish Flaxseed Semolina Squares 辣根亞麻籽粗麥方塊麵包
122 Seaweed Lemon Pain de Mie 海藻檸檬絲吐司
124 Whole Wheat Pain de Mie 全麥吐司

126 Glossary of baking terms 烘焙術語彙編
128 Acknowledgements 鳴謝

How does it work? 麵包是怎樣做成的？

Mixing and Kneading
攪拌與揉麵
p16-19
Hydration 水合作用
p20
Shaping 整型
p22-23
Proofing 最後發酵
p24
Enjoying 享用
p28-29

Your flour

Going to the supermarket and hunting for suitable flour for baking bread can be a real headache with so many different names and different brands. The quick reference table below will help you find the most suitable flour for the recipes in this book. It is worth noting that the ash content is not related to the protein level, for example, typical pasta flour (type 00) can be very strong in protein. Additionally, supermarket's whole wheat flour and rye flour are not divided in so many kinds of milling and thus easier to select.

In this book 書中	Common name 常用名字	Feature / Usage 特徵 / 用途
Flour type 45 45 號麵粉	Plain white flour 低筋麵粉	Lower protein content, cake, cookies and soft breads 低蛋白質，適合做蛋糕、餅乾和軟麵包
Flour type 55 55 號麵粉	In recipes, can be used for both type 45 or 65 中筋麵粉在食譜內，可替代低筋和高筋麵粉	
Flour type 65 65 號麵粉	Strong white flour 高筋麵粉	Higher protein, crusty bread, stronger gluten bread 高蛋白質，適合做外層香脆、筋性強的麵包
Flour type 00 00 號麵粉	Flour type 00 麵粉型號 00	Very fine flour, good protein, pasta, pizza 非常精緻麵粉，蛋白質良好，適合做麵條、披薩

The type of wheat flour is mainly defined by its ash content or by its extraction level. The extraction level is the quantity of flour obtained after the milling of 100 kilos of clean grain. This is also called the milling yield. The ash content is measured by the minerals left after baking flour at 900°C for about 2 hours. Minerals won't be combusted; thus, the miller can measure the ash content. Different countries use diverse ways of measurement for wheat flour. Most countries have specific law governing specification of flours and their designation.

Ash 灰份	Protein 蛋白質	Flour type 麵粉類別		
		U.S.A 美國	France 法國	Euro 歐洲
~0.4%	~9%	pastry flour 低筋麵粉	45	450
~0.55%	~11%	all-purpose flour 中筋麵粉	55	550
~0.8%	~14%	high gluten flour 高筋麵粉	80	800
~1%	~15%	first clear flour 特高筋麵粉	110	1100
>1.5	~13%	white whole wheat 白全麥麵粉	150	1500

In the above table we can observe the difference between different regions. For example, the US uses the ash and protein content to define the type of flour; European countries are roughly the same calculating on 100 grams of dry matters while French are calculating it over 10 grams. You will discover many types of flour along the recipes of this book; however, there are 3 main types widely used in baking. They are the white wheat flour, the whole wheat flour and the rye flour.

麵粉

踏進超級市場，有各種各樣的麵粉，名字和品牌各有不同，要找適合做麵包的麵粉無從入手，真傷腦筋。左頁的速查表，能幫助你找到最適合書中食譜的麵粉。值得一提，麵粉的麩質含量與蛋白質沒有關係，例如典型的麵條麵粉（型號是 00）含非常豐富的蛋白質。此外，超級市場的全麥麵粉和粿麥麵粉就沒有劃分太多磨製成品類別，反倒容易挑選。

小麥麵粉的類型由麩質含量或被抽出來的可用成份而決定。以每 100 公斤穀粒計算，可被抽取的麵粉量，所得份量稱為磨製粉的生產量。麥麩含量從烘焙麵粉經用 900℃烘烤 2 小時，餘下的礦物質所計算。透過未被燃燒的礦物質，磨坊可計算出麩質含量。由於不同國家的計算方法各異，大部份國家均有指定法律管制不同種類的小麥麵粉。

根據上表顯示，得知美國、法國和歐洲麥粉的差異性，例如：美國依據麩質和蛋白比例劃分麵粉類別，歐洲國家與美國相若，以每 100克為單位，粗略計算乾燥物，至於法國則以 10 克以上計算。你會發現書中食譜有多種麵粉，最為常用的只有 3 種，這就是白小麥麵粉、全麥麵粉和粿麥麵粉。

Leavening methods

Dry yeast

There are two main kinds of dry yeast. They are active dry yeast and instant dry yeast. Active dry yeast must be dissolved in water at 38°C and rest for a period of 15 minutes in order to have efficient results. If the temperature is much lower, for example at 20°C, the fermenting power will be reduced by about 35%.

Instant dry yeast is the most popular found in supermarket. It is important to mix the instant dry yeast at the beginning of the mixing in order to have all the granules of yeast dissolved properly. A too short time of mixing could leave some granules not dissolved. The recipes of this book are using instant dry yeast making the recipes home-baking friendly.

In terms of equivalence between fresh yeast and dry yeast, the general rule is that 10 gm of dry yeast is equal to about 35 gm of fresh yeast. Dry yeast is basically dehydrated fresh yeast; therefore, the main difference is the extraction of water from the fresh yeast. In terms of fermenting effect, they are equal. But the so called 'purist' bakers might say that fresh yeast gives better aromas.

Fresh yeast

Fresh baker's yeast is a cultivated unicellular mushroom measuring about 5 mm. Yeast is creating the fermentation of substances like sugar, fruit juices or raisins. With oxygen, yeast is multiplying. In doughs for example, yeast ferments and produces carbon dioxide (CO_2), causing the dough to rise. The biological name of yeast is: Saccharomyces cerevisiae. There are many kinds of wild yeast in nature and when we make a sourdough, for example, we are using these wild yeasts.

Yeast dies at around 50°C and is very sensitive to thermal differences. Yeast is composed of water (75%), proteins (12%) and carbohydrates (13%). Fresh yeast can be found pressed in small blocks or can be purchased in bulk. It can be kept for about 2 to 3 weeks at 5°C, and for logistic or stock control purpose, it can be frozen to up to 6 months; however, the effect of the yeast will decrease. It is important to defrost the yeast slowly and to use it immediately once thawed, also, after thawing, the yeast may become liquified.

Poolish

The rule of thumb for poolish (or sponge) is an equal part of flour and water mixed with a little yeast. It will proof over a relatively long period of time in a plastic container. The fermentation time can be up to 24 hours in a fridge or fewer hours at room temperature. The poolish will start losing its power once all the sugars present in the dough are consumed by ferment.

While making poolish, ensure the flour is well mixed to avoid lumps of flour. Also, it is important not to add salt in a poolish as it will reduce greatly the fermentation all together. The poolish brings aromas and a longer period of conservation to the bread. As well, the crumb color will be cream rather than white.

Sourdough

Real bread starts with sourdough - says the purist bakers. Sourdough is the exemplary natural method for leavening bread. Sourdough gives to the bread a spectacular crust as well as a very pleasant crumb. Breads will stay fresh for a longer period compared to conventional leavening methods. This method also prevents bread stalling with its natural bacteria. The slight acidic flavor of the sourdough gives the bread a great balance in taste. Each baker develops his or her own method and recipe according to the needs of the production. A good sourdough can be refreshed and maintained over many years.

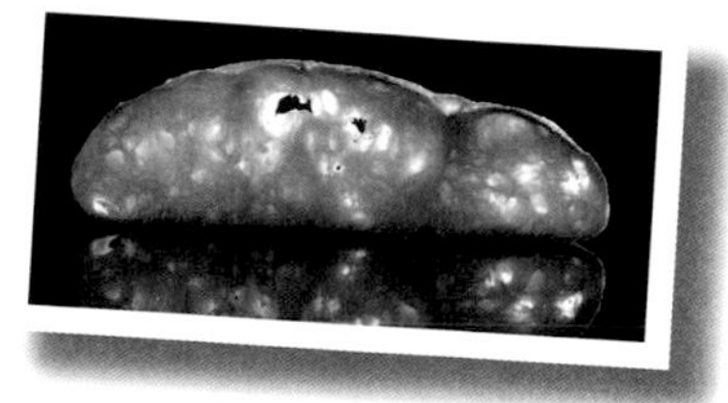

發酵媒介

乾酵母

這裏有兩種主要乾酵母：活性乾酵母和快速乾酵母。活性乾酵母必須溶於 38℃的溫水，停放 15 分鐘後，效果較理想。如果水的溫度較低，例如水溫只有 20℃，其發酵能力會降至約 35%。

快速乾酵母在超級市場內最常找到。它在物料融合前便要混和，確保所有酵母小粒子正常溶化。混合時間太短，會使部份粒子殘留而未能溶化。書中食譜採用這款容易使用的快速乾酵母，方便家庭式烘焙之用。

新鮮酵母與乾酵母的配份比例，按照換算準則是每 10 克乾酵母相等於 35 克新鮮酵母。乾酵母是新鮮酵母脫水後的製品，主要分別在於從新鮮酵母抽出的水份。就發酵效果而言，它們是一樣的，但所謂完美主義的麵包師傅或許會說使用新鮮酵母，更能帶出其芬芳香味。

新鮮酵母

新鮮烘焙酵母是 5 微米單細胞培植菌。它能製造出發酵物質如糖、果汁或葡萄乾。當與氧氣相遇，酵母會繁殖，例如在麵糰裏，酵母發酵會產生二氧化碳，導致麵糰脹大，酵母的生物名稱為"Saccharomyces cerevisiae"。在自然界有許多野生的酵母，當我們製造麵種時，就要用上野生的酵母了。

酵母對不同熱度很敏感，置於溫度約 50℃以上會死掉。它由 75% 水份、12% 蛋白質和 13% 碳水化合物（即醣）組成。新鮮酵母能被壓成細小塊或可供大量購買。在溫度約 5℃下，可保存 2~3 星期。就物流或倉存控制為目的，可置冰庫貯藏約 6 個月，但其效力會逐漸降低。值得一提，酵母要慢慢解凍，解凍後就必須立即使用，因為解凍後的酵母會化成液體。

速成麵種

速成麵種首要規則是用同等份量的麵粉和清水，再混合少許酵母，放進保鮮盒作長時間發酵。在冰箱的發酵時間 24 小時，而置於室溫則只需數小時。當所有糖份已進入麵糰協助發酵，麵種將開始喪失能力。

速成麵種完成時，確保麵粉完全混合而沒有粉塊。切記不要在麵糰放鹽，它會令所有材料的發酵力降低。發酵麵糰能帶出芬芳香味，延長麵包的保存時間。同時麵包層的顏色呈奶白。

老麵種

真正麵包由麵種為開端 —— 這是完美主義的烘焙師說的。麵種是發酵麵包的典型並為最天然的方法，它能令麵包外層特別香脆和有討喜的麵包屑。相對於傳統的發酵方法，麵包的保鮮期較長。這種方法能使麵包與天然細菌隔離，麵種的微酸味道可平衡麵包的食味，每個麵包師傅會因應產品需求而發展其秘方，一個好的麵種能不斷更新且保存多年。

Mixing and Kneading

It is important to measure and weigh each ingredient before mixing any dough. The mixing of ingredients must be done in slow speed in order for all the components to blend properly; this is the homogenization process. When the speed is increased, under the mechanical action, the glutenin and gliadin will form a network of gluten. The friction of starch cells will allow them to absorb more humidity as the temperature increases. This particular friction process is being studied and analyzed as dough rheology, an essential factor in bread making. When increasing kneading speed, there should be no more traces of water or flour. The baker should now touch the dough to feel if the dough is too soft or too hard and make any adjustment if needed. If any flour or water is added, it is important to adjust it early enough to allow the starches to be transformed adequately.

A common process to start kneading bread dough is called autolyse. It's a process you can apply to many recipes by only mixing the water and the flour of the recipe and let it rest between 20 minutes to one hour. It will allow the flour to start hydrating and the enzymes, such as protease will start to break down proteins in the flour. The autolyse will make the dough stronger and extensible. Because of the early hydration of starch, the dough will have better stretching properties. While the glutenin and gliadin starts to be broken down, the extensibility of the dough will increase. This process needs to be done just by mixing both ingredients shortly, without stirring for a long time.

In flour, the absorption level of humidity is determined by different factors:

- Moisture content (lower moisture = higher absorption)
- Protein content (lower protein = lower absorption)
- Flour grade, ash or bran content (higher value = higher absorption)
- Water soluble protein (higher value = higher absorption)
- Damaged starch (higher value = higher absorption)
- Enzymes activity (amylase) (greater activity gives lower absorption)

The salt is generally added at the end of kneading in order to allow the gluten network to fully form. Salt will tighten the dough with its hygroscopic properties. If butter is needed, it has to be added at the end of mixing to allow the gluten network to form completely. Butter will create an insulating layer between the cells of starch and will decrease the formation of gluten. Before the end of kneading, the baker should check the elasticity of the dough by pulling a small part of dough between his hands. For proper gluten, you should be able to almost see through the dough without breaking it. If it breaks, it needs more kneading. Dough with whole meal flours will have less elastic textures due to the coarse milling.

Bread dough has a peak point of optimal kneading. The baker always stops the kneading before the peak point of the extensibility of the dough. If it passes the peak point it will decrease and the dough will be over kneaded. When trying to pull the dough, it would tear into small parts or if the dough contains eggs, sugar and butter, it will create very long strings. In this case, we call the dough 'burnt'.

The recipes in this book primarily use spiral mixer and single arm mixer. For smaller dough we also used planetary mixer. A planetary mixer is equivalent to most mixers present at home. The bowl of the spiral mixer moves on the opposite way of the arm movement to increase friction, while in the planetary, the bowl is fixed. The baker himself will judge if the kneading is proper or not. It will vary according to the type of mixer, flour, weather and temperatures.

The following mixers are the most common in small to medium size bakeries.
以下是中小型麵包店常用的攪拌器：

Mixer 攪拌器	RPM 每分鐘轉數	Class 類別	Advantage 優點	Disadvantage 缺點
Single Arm 單臂式	28 to 60	Conventional 傳統式	Good kneading 揉麵能力良好	Kneading takes long time 揉麵時間長
Twin Arm 雙臂式	28 to 60	Conventional 傳統式	Higher volume of oxygen 氧氣容量高	Strong flour dust 麵粉塵較多
Planetary 星球式／軌道式	Adjustable 可調校	Mixed 混合式	Great flexibility of speed 變速能力佳	Limited sizes 尺碼有限
Spiral 螺旋式	80 to 120	Intensive 密集式	Excellent kneading process 揉麵過程極佳	Higher risk of over-mixing 有過度攪拌的危機
Fork 分叉式	80 to 120	Intensive 密集式	Higher volume of oxygen 氧氣容量高	Higher risk of over-mixing 有過度攪拌的危機
Double Cone 雙筒式	80 to 140	Intensive 密集式	Dough is handled carefully 能細緻處理麵糰	Only for large amount of dough 只能處理大量麵糰

Temperatures of dough

Temperature of dough is very important in baking. It determines the length of resting and proofing time. Also, the baker will adjust the temperature of the dough according to its production plan and leavening method. As a general rule, the range of temperature for a cold dough kneading is between 18°C and 22°C and for warm dough kneading between 26°C and 30°C. Higher temperature increases the swelling ability of the starch molecules, especially for whole meal flours. Excessive warm dough will proof too fast and not meet the proper development of flavors or required qualities; on the opposite, very cold dough will not proof and not develop to the necessary standards.

The basic temperature, present in all recipes where the dough is kneaded, follows a simple calculation method. It has for purpose to calculate the temperature of the water needed.
基本溫度借用算術程式來計算出書中食譜的攪拌時間。為了達到計算目的，當混合麵糰時，便需要計算出水溫。

Example: 例子

Dough temperature target 麵糰的目標溫度	27℃
- Dough friction heating 麵糰摩擦熱度	6℃
= Target 目標	21℃
Multiplier 3x21℃ = 倍增	63℃
- Room temperature 室溫	25℃
- Flour temperature 麵粉溫度	24℃
= Water temperature 水溫	14℃

In some bakeries, there is a cooling water machine that helps the baker to have a steady supply of cooled water. However, in warm climate, cold water is difficult to obtain without machine. Therefore, the use of ice cubes or ice chips is recommended following a simple calculation assuming that the temperature of the ice is of -0.5°C.

Example: 例子

Temperature of available water 可用水的溫度	20℃
- Needed water temperature for recipe 食譜所需水溫	14℃
= Difference 相差	6℃

Therefore the ice ratio will be: 冰量比率
6% of ice (-0.5℃) and 94% of water at 20℃ = 100% of the liquid needed in the recipe.
6% 冰粒（-0.5℃）和 94% 水溫 20℃ =100% 食譜所需液體。

攪拌與揉麵

攪拌麵糰前需要計算和量好各材料，這是非常重要。使用慢速，徹底攪拌材料，以求各材料正常融合，這是均質化過程。當速度增加，在機械運作下，麥穀蛋白和麥醇溶蛋白會構成麵筋網絡。澱粉細胞摩擦時，溫度升高，容許吸入更多濕度。經研究和分析這特別的摩擦過程—— 麵糰流變學，它是麵包製造的要素。當搓揉麵包的速度增加，便不會有水或麵粉的痕跡可尋。麵包師傅在此時應該觸摸麵糰，感受它是否太軟或太硬，作出調節。任何麵粉或清水添加於麵糰內，就要早些調足所需，好讓澱粉有足夠時間轉化。

自解法是指開始攪拌麵糰時，一個慣常用的程序。此過程可應用於許多食譜中需要把清水和麵粉混合，讓其靜止 20 分鐘至 1 小時，待麵粉脫水發酵，正如蛋白酵素類物質會開始破壞麵粉的蛋白質。自解法會令麵糰強壯和易於擴張，因為澱粉質早點脫水，麵糰的伸展屬性會比較好。當麥穀蛋白和麥醇溶蛋白開始被破壞，麵糰的擴展度會增加。
這過程需要在短時間內把兩種材料混合，不能作長時間攪拌。

麵粉裏，其吸濕程度取決於不同因素：
　麵粉的濕潤度（低濕潤度 = 高吸濕能力）
　蛋白成份（低蛋白質 = 低吸濕能力）
　麵粉等級、灰質或麥糠成份（高價值 = 高吸濕能力）
　水溶性蛋白（高數值 = 高吸濕能力）
　破損澱粉（高數值 = 高吸濕能力）
　酵素活性（澱粉酵素）（活性越高，吸濕能力越低）

在最後揉麵過程才添加鹽，可徹底地架起麵筋網絡，且它與吸濕物質可拉緊麵糰。如果必要加奶油，就在最後攪拌時才添加，令麵筋網絡完全建立。奶油會引起澱粉細胞和夾層分離，將會降低麵筋的形成。待最後攪拌之前，麵包師傅應該用手把小部份麵糰拉扯檢查它的彈性。正常的麵筋，可看到網絡呈透明而不破損。如果它破裂，就要繼續搓揉。由於小麥的磨研粗糙，其彈性質地會較弱。

麵糰有最理想的揉麵極限。麵包師傅應在揉麵極限前，停止揉麵。如果超越了極限，會令麵糰過份揉麵。拉開時會被撕成小塊；又因它含有雞蛋、糖和奶油，將會創造長纖維，在這情況下，我們稱為麵糰「燒焦 / 過度攪拌」。

書中食譜主要用螺旋形攪拌器或單臂式攪拌器。少量麵糰可用球形攪拌器，為家庭式常用攪拌器。螺旋形攪拌器的攪拌缸與其臂的移動，背道而馳，意即其臂運行而攪拌缸固定，增加摩擦力。麵包師傅可根據攪拌的運作情況，自行調節，因不同型號的攪拌器、麵粉、天氣和溫度會帶來變數。

麵包店常用攪拌器詳見第 17 頁的表。

麵糰的溫度

在烘焙時，麵糰溫度非常重要。它決定了靜止和發酵時間，同時麵包師傅會因應生產計畫和發酵方法，調節麵糰溫度。一般規則，冷凍麵糰的攪拌溫度由 18℃至 22℃；而一般麵糰的攪拌溫度是 26℃至 30℃。高溫會令澱粉微粒子增加膨脹能力，特別是全麥麵粉。溫度過高的麵糰會膨脹很快，味道和所需的品質未能正常發展。相反地，溫度過低的麵糰就無法膨脹，未能發展到所需標準。

有些烘焙工場，備有冷水機器給麵包師傅供應穩定冷水。然而，在溫暖的氣候，沒有機器的協助，很難取得冷水，為了解決這問題，建議使用碎冰或冰塊調節水溫，假設碎冰溫度為 -0.5℃，簡單計算方法如下：

Hydration

The structure of the crumb, the inner part of your bread, is defined by the level of hydration. The hydration level is the amount of water used in your recipe, and as with other ingredients in your bread, the hydration ratio is calculated as a percentage to the total weight of the flour. In typical bread it ranges from 50% to 90%. Some specialties will reach higher hydration level and some even lower levels. While water is the lead indicator of hydration in dough, milk or eggs are also contributing to the hydration level in dough.

The hydration process starts at mixing, during the homogenization stage and further continues when the kneading starts as explained in the mixing and kneading chapter.

Below are the crumbs and their different textures according to hydration levels.

水合作用

麵包內碎屑的架構為水合作用層。這水合作用層是食譜中的用量水和產品的其他用作製造麵包的材料相若，水合作用的百分率以麵粉總重計算出來。麵包可達 50% 至 90% 不等。有些特別情況會出現較高水合作用層或較低的水合作用層。主導麵糰的水合作用的指標，而鮮奶或雞蛋更有助其發展。

水合作用過程從攪拌開始，在均質化時期和麵糰進一步發展的變化，已於攪拌麵糰的章節解釋了。

根據不同水合作用層，可看到麵包內部碎屑及不同程度的質感在演變中，參照上圖。

Fermentation

The goal of the fermentation process in bread making is to develop the dough texture and to influence its taste. Factors affecting the fermentation are the flour strength, enzymatic activity of the flour, the leavening method used and the desired final product.

During the fermentation, the yeast converts the fermentable sugars into CO2 and ethanol. The CO2 is then released into gas cells and the dough expands. Lactic acid and acetic acids are generated. Within the first 25 minutes of the fermentation, the pH typically drops from 6.0 to 5.0.

Any fat contained in dough will help creating an extra layer in the bubble network and will help to retain the fermentation gases under a controlled proofing temperature below the melting point of the fat used, for instance butter is about 33°C.

Bulk fermentation

After the mixing and kneading of the dough, the bulk fermentation happens either in the mixing bowl itself or in a plastic container covered with a plastic sheet. Time of bulk fermentation varies according to each recipe and method, but the longer the better, except for some specialty bread requiring an earlier shaping. During the bulk fermentation the dough starts to produce CO2 and gain in taste and texture.

發酵

製作麵包時，發酵過程的目的是發展麵糰質地和改變其食味。影響發酵的因素有麵粉本身的溫度、酵素活性、發酵方法和要求製品的結果。

在發酵時期，酵母會由醣轉化為二氧化碳和乙醇，而二氧化碳會釋出氣體和令麵糰膨脹和伸展，繼而產生乳酸和醋酸。在最初發酵的 25 分鐘裏，酸鹼值一般會從 6.0 降至 5.0。

任何在麵糰的油脂會幫助氣泡網絡產生額外層，並在受控的發酵溫度下保留令油脂達熔點，例如奶油的熔點大概是 33℃。

基本發酵

麵糰經混合和揉搓後，會在攪拌缸內或已蓋保鮮膜的塑膠器皿進行初次發酵。發酵時間因應個別食譜和方法而變更，時間越久越好，除了某些特定麵包需要提早整型。在基本發酵過程中，麵糰產生二氧化碳，味道和質感便在這時形成。

Resting

The resting time depends on recipes and allows the dough to rest before the shaping. In general, the dough should not be shaped right after being divided; it will often be pre-shaped for proper resting on the table. This time of rest allows the dough to relax and not being over stretched. Remember to always cover your dough while resting or else the dough envelop will dry and form a crust.

鬆弛或醒麵

麵糰在整型前需要鬆弛麵筋，時間按不同食譜而異。一般情況，麵糰因分割後不立即整型，所以會預先整型，然後置於桌上作鬆弛程序。因這時可讓麵糰鬆弛但不會過度伸展而失去筋度。切記麵糰在鬆身前應蓋上保鮮膜，否則它的表面會變乾而呈龜裂狀。

Shaping

Weighing and dividing

Weighing and dividing can be done by hand or by machine. While dividing the dough by hand it is important not to pull on the dough and not to tear the dough. Also, if it is divided by hand, the baker should try to weigh homogeneous pieces of dough, as close as possible of the targeted weight and avoid adding several small pieces together.

Shaping methods

There are unlimited ways to shape bread. Each region in each country has its own specialty and shape. The basic shapes are the oval loaf and the round loaf. The baguette is a derived of the oval loaf, stretched in length. Each shape can then be done in any sizes. Shaping is most likely the exercise that the baker needs to practice the most in order to find his own style. Although there are some basic movements and principles in shaping, individuals tend to develop their own habits after a while.

Key points during shaping

There are several factors to take into account while shaping bread. The table you will work on is one of the key elements in the ability to shape bread correctly. Once declared not proper to use by some HACCP bodies, there was controversial studies showing that natural bacteria in wood were having a positive impact rather than none present in plastic boards, being without natural defenses. Therefore, I am recommending to use a wooden table for different reasons, one of which being the pressure created by the baker on the table that is partially absorbed by the wood. On a long term prospect, if the baker shapes on a marble or stainless steel table, the shock absorption by the body will be harmful to arms and shoulders. The height in which you work is also very important for your back.

The hand pressure on dough needs to be adapted to the type of dough and shaping. Depending on the dough power and leavening method, the pressure needs to be adapted for a tight shaping or a relaxed shaping. This will influence the texture and shape of the bread once baked. During the shaping, it is important to burst large gas bubbles to prevent having a few very large holes in your baked bread.

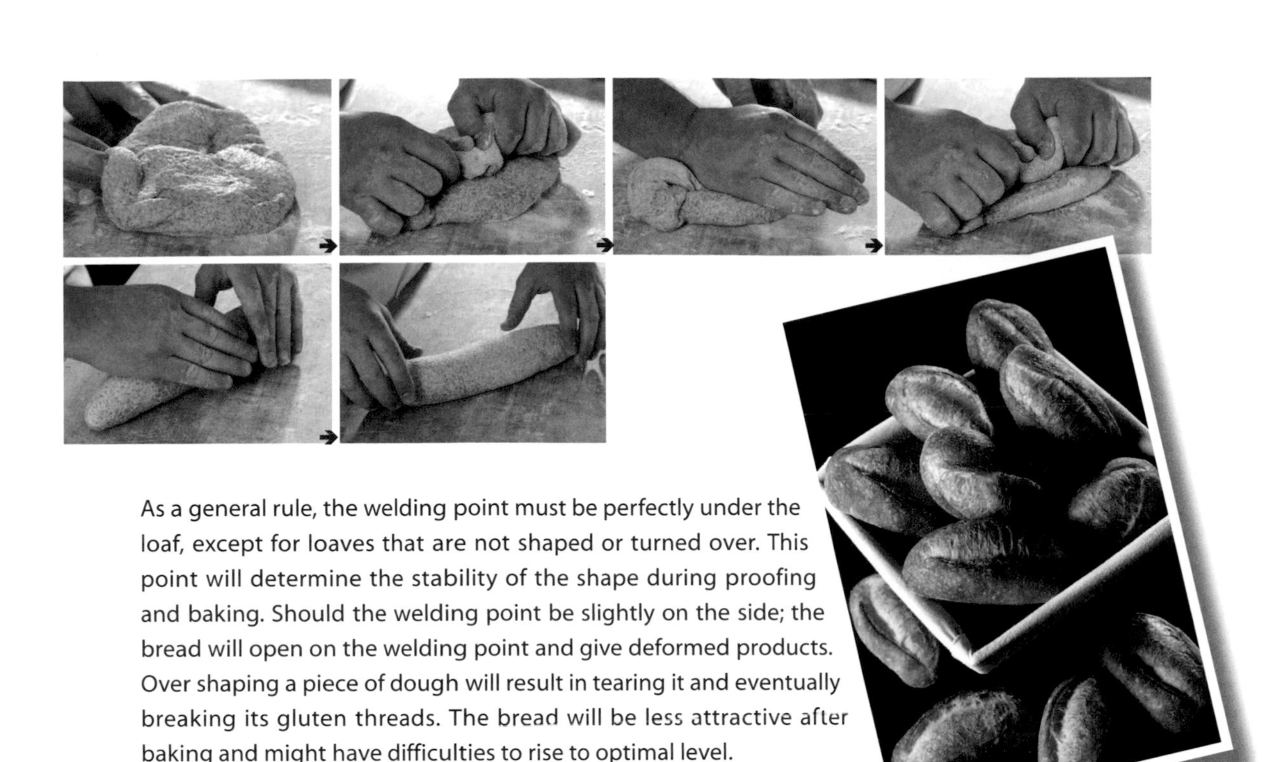

As a general rule, the welding point must be perfectly under the loaf, except for loaves that are not shaped or turned over. This point will determine the stability of the shape during proofing and baking. Should the welding point be slightly on the side; the bread will open on the welding point and give deformed products. Over shaping a piece of dough will result in tearing it and eventually breaking its gluten threads. The bread will be less attractive after baking and might have difficulties to rise to optimal level.

整型

量重量和分割

量重和麵糰分割可用手或機器處理。用手把麵糰分割，最重要是不能拉扯和撕裂麵糰。同時，如果用手分開，麵包師傅應嘗試把每個麵糰均一，盡可能接近所訂定的目標重量，以及避免數小塊疊在一起。

整型方法

這裏有無限方法替麵包整型。每個國家的省份均有其獨特之處。最基本的造型是橄欖（蛋形）和圓形。法國長棍麵包便是把蛋形麵糰伸展長度而成的橄欖形。每個造型可做出不同尺寸，麵包師傅常常練習麵包整型，才可以創出個人風格。儘管整型有很多基本手法和原則供參考，個別麵包師傅經練習一段時間後，就會訂出自己習性的麵包造型了。

整型訣竅

麵包整型時，有許多因素需要考慮，工作檯面是正確整型的主要元素。有些食品安全認證組織曾告誡不正確使用工作檯面，多項具爭議性的研究指出含天然細菌的木製檯面，比沒有天然細菌的塑膠檯面整型有正面影響，反而塑膠檯面缺乏自然抗力。所以，我建議使用木製檯面的原因。它可卸除因麵包師傅搓揉麵糰所產生的壓力。長遠而言，如麵包師傅在大理石或不銹鋼檯面長時間工作，吸取了許多反震力，令手臂和肩胛受損。至於工作檯的高度更是麵包師傅的背部受傷的關鍵。

究竟要用多少力度壓麵糰才適合？因應麵糰類別和整型而決定。換句話說，這有賴麵糰韌度或發酵方法，才按照麵包整型的鬆緊度而採用適當壓力。壓力是否施用得恰當會直接影響到烘焙製品的質感和形態。整型時，必須把麵糰的空氣壓出，否則烘焙後的麵糰會出現大氣孔，不夠細緻。

一般而言，麵糰的接合處必須置於麵包底部，除了那些不用整型或翻轉的麵包類，因為這點決定麵糰經發酵和烘焙後的穩定性。如麵糰接合處須偏在一邊，處理不當，麵包將會出現爆裂或造成畸型。如麵糰過度整型，最終出現撕裂或弄斷麵筋。烘焙後賣相欠理想，及難於上升至最佳水平。

Proofing

Once the dough is shaped, the baker might have decided to conduct its dough directly in the proofer or he might have decided to have a slow fermentation. There are special rooms with controlled atmosphere to control the fermentation over a relatively long period of time. However, at home, you can use your fridge as a replacement. Not optimal, but it works. Or else, the baker might have decided to conduct its dough on a long fermentation at room temperature, covered under a plastic sheet to avoid crusting.

The final proofing is the last step before the actual baking. The baker's senses are sole judges on whether or not the time is right to score and bake a loaf. An important point of the final proofing is allowing the crust to dry a little in order to do clean scores. If the outer layer of the bread is too wet, the blade will stick to the dough while cutting.

The proofing times indicated in this book are for reference. It will depend largely on the temperature and humidity level as well as the size and kind of bread you produce.

最後發酵

整型後，麵包師傅按麵糰情況而決定直接送進發酵爐或緩慢發酵。這裏有控制氣壓的特別房間，以操控相對長時間的發酵過程。在家裏只有冰箱取代發酵箱，雖不理想，但仍可使用，或是麵包師傅可把它蓋上保鮮膜，以防止麵糰變硬，然後放進大保鮮盒而放置室溫下作長時期發酵。

真正把麵糰入爐前的最後步驟便是最後發酵。麵包師傅憑個人感覺和判斷該何時在麵包劃紋繼而進行烘焙。在最後發酵時，重點是容許有少許麵包皮外層略乾，以便留下清晰劃紋。如果麵糰的外層過濕，割紋時，刮刀會黏貼在麵糰不夠俐落。

書中的最後發酵時間只作參考，製作時間取決於其溫度、濕度和大小和種類而進行。

Scoring

There are different types of blades such as razor blades, special bread blades or other very sharp knifes used to score bread. The angle of the blade, the deepness of scoring and the amount of scoring are depending on the proofing level and on the type of dough the baker has done. Mostly, it is the baker that judges these factors while looking and touching the loaves before scoring. When scoring, the gesture of the baker must be well thought and confident; a score must be done in one stroke. Before scoring, ensure to lift slightly each loaf to make sure that they are not stuck on the fermenting cloth. When placing the loaves on the cloth, allow enough space for the bread to expand and to have a uniform coloration.

刻紋

把麵包刻紋可用不同類型的剃刀，如刮鬍刀片、特製的麵包刀片或其他利刀劃紋。麵包師傅下刀的角度、劃紋深度和數量，需按照發酵度和麵糰類別而為。一般而言，麵包師傅會先觀察和觸摸麵糰後才作決定。劃紋時，麵包師傅定必已準備站姿，然後充滿信心兼俐落下刀。在劃紋前，確保每個麵糰微微分離，不會黏貼在發酵布。當麵糰放在布上，容許有足夠空間讓麵包伸展直至顏色一致。

Baking

Types of oven and temperatures

The first step of a successful baking is to prepare the oven at the right temperature ahead of time. Many different types of oven are available on the market. To obtain a better quality loaf, an oven with a stone bed is recommended. The direct contact of the dough on the stone gives direct heat transfer to the bread and the product will develop all of its aromas, a great crust, a good volume and an attractive shape. The heat balance is the distribution of the heat within the oven. Most of the time, by natural heat loss, the temperature near the doors of the oven is lower, therefore the baker must move the loaves around during baking to obtain a homogeneous coloration of the dextrin. The temperatures and the baking times indicated in this book may vary; it will depend largely on the type of oven you use and on the quantity of bread you produce.

The exhaust function

Typically, most commercial ovens you have at home are not equipped with an exhaust function. The functionality of the exhaust is simply to keep steam inside the oven or to release it through a valve.

For home baking, it is a common practice to spray your loaf with water before baking. Spraying directly in the oven might not be the best for your oven's duration. Another way to bake is to preheat a stoneware pot deep enough to accommodate your loaf, start the baking of your loaf in it to later finish proper coloration outside the pot. Extreme home bakers using stone plates will flip a metal container over their loaf and spray steam in it.

Baking stone

Using a stone in home baking makes a big difference to your end product, whether you bake a pizza or a loaf of bread. The stone can be the same nature as heat resistant "engineered" brick (used to build wood fire oven). Another common natural stone used for baking is cordierite. Granite and slate are also fine; of course any stone used to bake should not be glazed.

A thickness of about half an inch will be appropriate. The thicker the stone, the more time it will take to heat up. If you use a gas oven, it should take a total of about 45 minutes to bring the stone to even heat. And if you are using an electrical oven, it will take about 60 minutes. In an oven with bottom and top heat control, the stone can be directly on the bottom of the oven, but if the heat is too strong from the bottom, you might need to keep your stone on a grid at the lower level in your oven to avoid a too intense heat.

Every stone, either man made or natural have a certain degree of moisture in them and you will need to "dry" them or else the stone will crack or break at first usage. So before baking with your stone, you have to warm it up gradually to "temper" it. For example 1 hour at 50°C followed by 1 hour at 100°C, followed by one hour at 150°C degrees and finally half hour at 200°C. There is no need to get an expensive stone with brand name, visit a local stone shop and ask them to cut a slab of your desired stone for the size of your oven.

To place your bread directly on the heated stone you will need a small wooden board or baker's peel. Choose a size that fits your stone. Simply dust your wood with fine semolina or flour to avoid sticking and place your proofed loaf on it and slide it confidently on the stone. You can use the same wooden board or peel to bring out your baked breads.

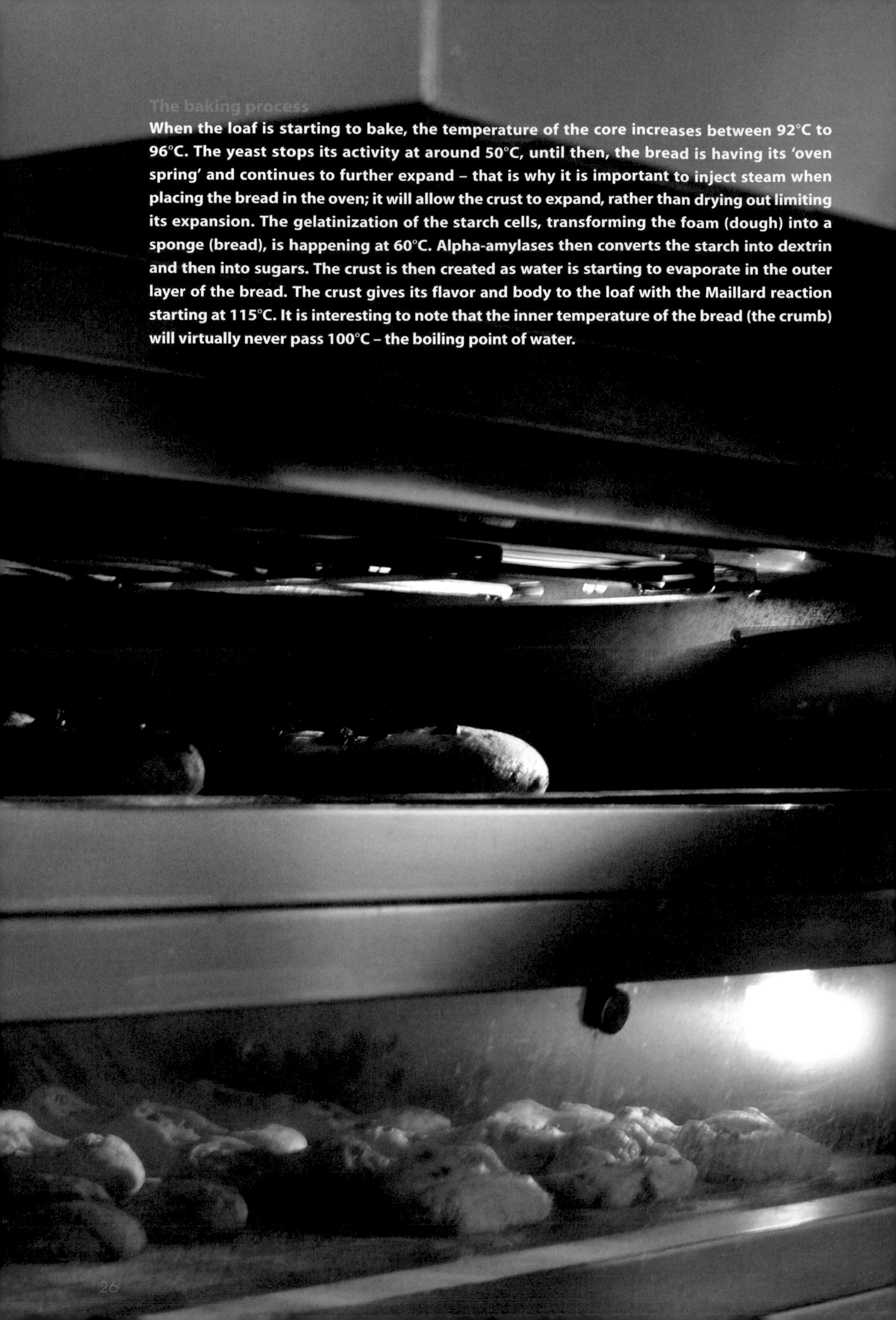

The baking process

When the loaf is starting to bake, the temperature of the core increases between 92°C to 96°C. The yeast stops its activity at around 50°C, until then, the bread is having its 'oven spring' and continues to further expand – that is why it is important to inject steam when placing the bread in the oven; it will allow the crust to expand, rather than drying out limiting its expansion. The gelatinization of the starch cells, transforming the foam (dough) into a sponge (bread), is happening at 60°C. Alpha-amylases then converts the starch into dextrin and then into sugars. The crust is then created as water is starting to evaporate in the outer layer of the bread. The crust gives its flavor and body to the loaf with the Maillard reaction starting at 115°C. It is interesting to note that the inner temperature of the bread (the crumb) will virtually never pass 100°C – the boiling point of water.

烘焙

烤箱類型和溫度

烘焙成功的第一步，預先調校正確的烤箱溫度。市面上有許多不同類型的烤箱，如要獲得優質麵包，建議選用有石床的烤箱。因為它能直接與麵糰接觸，熱力直透產品，能幫助其產生香味、優質鬆脆的外皮、合適體積和引人的賣相。熱量平衡是指烤箱的熱力均勻。許多時侯，熱力會自然消失，因接近爐門的溫度較低，所以麵包師傅必會把麵包移動，務求各麵包的糊精同質上色，達致色澤均一。書中所指的爐溫和烘焙時間會因應烤箱型號和入爐的麵包數量而調節。

排氣功能

許多在家使用的商業用烤箱是沒有排氣功能。這功能讓蒸氣保存於爐內或可借活門釋放熱氣。

在家烘焙時，最慣常的做法是在入爐前於麵糰噴水。而在爐內直接噴水並不是最好的方法，因會縮減烤箱的壽命。另一烘焙方法，預先燒熱粗陶器壺，並確保有足夠深度容納麵糰，開始烘焙，讓麵糰置於壺內，直至最後階段，把麵包放在外面，出現所需顏色。狂熱的家庭烘焙師會把金屬器皿翻轉於石碟上，蓋上麵包，再噴上蒸氣。

烘焙石

在家使用烘焙石，不論你烘焙披薩或麵包，能令產品有很大的不同。該石可以跟抗熱功能的耐火磚（建造營火烤爐的那種）同類便可。此外，較慣常用的石是堇青石。花崗石和板岩都不錯，當然無論你用任何石材，都要沒有上釉那些。

厚度約半吋是最適合。石頭越厚，預熱時間也越長。如用煤氣烤爐，需要用 45 分鐘加熱，直至熱力均勻。要是用電烤箱，就要 60 分鐘了。如烤箱的底火和上火，烘焙石能直接置於爐底，底部熱力過強，你需要把石頭放在架上並置於烤爐的最底部，避免熱力過於集中。

每一塊石，無論人造或天然的，都擁有某程度的濕度，你需要讓它慢慢變乾，否則首次用後會呈現裂紋，甚至斷裂。所以烘焙你的石頭之前，你需要漸進式把它加熱。例如，1 小時達至 50℃，接下來的 1 小時至 100℃，隨後 1 小時後至 150℃，最後半小時達 200℃。無需購買有品牌又昂貴的烘焙石，只要到本地石材店，要求他們按家中的烤箱尺碼裁出所需石塊便可。

把麵包直接放入已燒熱的石塊，需要用小木板或爐架。選一塊與烘焙石尺碼相當的，撒少許米粉或麵粉避免黏底，然後自信地把已發酵的麵包滑入石上，待烘焙完成後，可用那塊木板或爐架取出麵包。

烘焙過程

當麵包開始烘焙，烤爐的中心溫度在 92℃至 96℃之間。酵母會在 50℃停止活動，其後，麵包會開始在烤爐內膨脹，這亦解釋了為何入爐後要注入蒸氣，因可阻止外層變乾，好讓麵包在 60℃時，澱粉的作用使泡沫（麵糰）變成海綿狀（麵包）。澱粉酵粒子把澱粉轉化為糊，再變作醣份，麵包香脆堅硬的外層便在這時形成，因外層的水份開始被蒸發掉。溫度達 115℃時，幫助麵包帶出香味。有趣的是，麵包中心的溫度反而永遠過不了 100℃（水的沸點）。

Enjoying

When the heavenly time of pulling your baked loaves out of the oven, the bread will need to breath in aerated basket or shelves. The level of humidity coming out of the bread straight after baking represents about 1.5% of its own weight. Therefore, if the bread is in closed basket, it will become soggy right away. Additionally, extreme difference of temperature should be avoided, i.e.: don't cool your hot bread in the fridge.

To control the quality of baked bread, we can measure its acidity by measuring the pH. On the pH scale, acidic will be 1 and alkaline will be 14; 7 being neutral. The value of sourdough dough is of 3.5 to 4.5 and regular bread dough without sourdough should be about 5.7. The temperature affects pH measurement, but temperature is electronically leveled using certain tools.

Freezing dough and bread

For logistical purpose or production planning, freezing raw dough or par baked bread is a possibility, however from an artisan point of view; bread should be produced fresh every day.

It is generally believed that -25°C is adequate to preserve the dough from any physical damages to the yeast cells. However, studies have shown that temperatures below -30°C reveal the real stability of dough. Beside the storage temperature, the major issues of frozen dough is the speed of cooling and the speed of thawing since both of them are related to the formation of ice crystal.

Ice crystals are the main factor affecting quality in dough freezing. The formation of ice will depend on the competition between the ice nucleation and the thawing that needs to allow the dough to relax back to a thermodynamically stable state. The main concern is the role of water during the freezing process. Naturally, a water molecule migrates toward the ice embryos. For example if a loaf of bread is fully baked and frozen, the crust will contain less water than the crumb, therefore there will be a minimal migration of water from the crust.

On the other hand, in the higher humidity environment of the crumb, under a long freezing process, the water crystals will form and expand, pushing the crust outward. Once thawing, the ice crystals of the crumb will melt and the crumb will be back to its original shape and the crust will be broken and fall into pieces from the crumb. Using a blast freezing facility guarantees a better freezing process.

享用

當麵包從烤箱拉出是最令人歡欣喜悅的時刻，麵包需要在麵包籃或架子上呼吸。濕度會立即從麵包滲出，約佔其重量的 1.5%。所以，如果麵包放在密封的籃子內，它會變得潮濕。除此之外，應該避免極大的溫差，不要把熱騰騰的麵包放入冰箱。

控制麵包的品質，我們可用酸鹼度計算其酸性，在酸鹼度表，酸度是以 1 代表，鹼則是 14，7 屬於中位數。麵種的酸鹼度是 3.5 至 4.5。正常沒有加入麵種的麵包大概是 5.7。溫度會影響我們計算酸鹼值，需一些工具才能電子化地量度。

麵糰和麵包的冷藏

從物流目的或生產計畫的目的而言，冷藏生麵糰或半生麵糰是可行的，但麵包師的觀點，麵包需要每天生產比較新鮮一點。

一般人認為 -25℃的溫度能保存麵糰不會損害酵母細胞，然而一些調查報告顯示，溫度降至 -30℃影響到麵糰的穩定性。除了貯存溫度，冷凍麵糰的重點是其冷凍速度和融化速度之間的相互關係，這兩者和水結晶的成因有密切關係。

冰晶是影響冷凍麵糰的主要因素。冰粒的形成視乎晶核及用以使麵糰回暖至鬆弛狀態的熱力穩定性。在冷凍過程，水扮演很重要的角色，水分子轉化為冰胚，例如麵包是完全烘烤和冷凍，其外層的水份含量便少於內層，所以這裏會有少量的水份從外層釋出。

另一方面，濕度較高的環境即麵包內層，經長時期冷藏過程下，水結晶會形成和膨脹，造成硬的外層。待融化，內層的冰晶溶化，便會回到原來的形態，麵包外層將會碎掉。利用急劇性的冷藏設施能確保有較好的冷藏過程。

Your baking guide
圖例解讀

Room Temperature（℃）
室溫（℃）

Water Temperature（℃）
開水溫度（℃）

Dough Temperature（℃）
麵糰溫度（℃）

Kneading Time （Spiral Mixer）
搓揉時間（螺旋形混合器）

Dough Temperature（℃）
麵糰溫度（℃）

Bulk Fermentation （hrs. at ℃）
麵糰發酵（指定小時於指定℃）

Weighing （gm.）
重量（克）

Rest Time （hrs./mins.）
醒麵時間（小時 / 分鐘）

Shaping （Loaf shape by hand）
整型（用手搓揉成指定形狀）

Final Fermentation （hrs.）
最後發酵（小時）

Initial Baking Temperature （℃）
最初烘焙溫度（℃）

Baking Temperature （℃）
烘焙溫度（℃）

Baking Time （mins.）
烘焙時間（分鐘）

A touch of sweetness

甜麵包的接觸

蜜餞薑粒皇家麵包

Candied Ginger Pan D'oro

CHEF'S TIPS This recipe calls for very little yeast, a very long fermentation and a very liquid dough making it very moist and delicious. Pan d'Oro is traditionally made for the end of year festivities.

麵包師建議 這食譜只需很少量的酵母，經長時間發酵和稀流麵糰，便可製造出非常濕潤和可口的麵包。蜜餞薑粒皇家麵包是傳統年末節慶而準備的麵包。

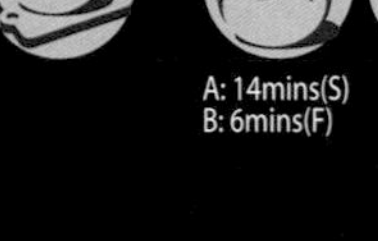
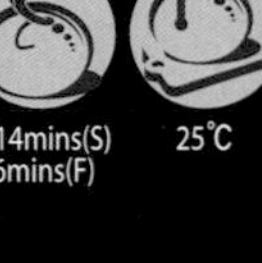

63℃ | A: 14mins(S) B: 6mins(F) | 25℃ | one night | 50 gm | 20 mins | round | 3 hrs | 180℃ | 180℃ | 25 mins

INGREDIENTS

	INGREDIENTS	RATIO
280 gm	Flour type 45	100%
80 ml	Milk	28.57%
90 gm	White sugar	32.14%
110 gm	Butter	39.29%
180 gm	Eggs	64.29%
8 gm	Dry yeast	2.86%
5 gm	Salt	1.79%
half pc	Lemon zest	
2 tbsp	Amaretto	
1 tbsp	Accacia Honey	
75 gm	Diced candied ginger	
	White sugar	
	Icing Sugar	

	材料	比例
280 克	低筋麵粉	100%
80 毫升	鮮奶	28.57%
90 克	細砂糖	32.14%
110 克	奶油	39.29%
180 克	雞蛋	64.29%
8 克	乾酵母	2.86%
5 克	鹽	1.79%
1/2 個	檸檬皮	
2 湯匙	杏仁甜酒	
1 湯匙	洋槐蜜	
75 克	蜜餞薑粒	
	細砂糖適量	
	糖霜適量	

METHOD

1. Knead the dough and add the salt 3 minutes before kneading ends.
2. Let it rest at room temperature for half an hour and fold it once.
3. Keep the dough in a covered plastic container overnight in the fridge at 5℃ .
4. The next day, fold the dough again, weigh it and pre-shape it in round shape.
5. Allow resting and shape the rolls. Place them in metal ring lined with paper.
6. Proof at room temperature for about 3 hours.
7. Brush water, dust sugar and icing sugar heavily and bake with exhaust open.

製 法

1. 搓揉麵糰，在接近完成揉麵的 3 分鐘前加入鹽拌勻。
2. 在室溫下靜置半小時，摺疊一次。
3. 把麵糰放在塑膠器皿，待 5℃冰箱過一夜。
4. 翌日，再次摺疊麵糰一次，量重量，初步做出圓形。
5. 讓其鬆弛和做成小卷麵包，放在已鋪烘焙紙的金屬圈。
6. 置室溫下醒麵 3 小時。
7. 刷上清水，撒上細砂糖和大量糖霜，放入已開啟排氣活門的烤爐烘焙。

皇冠麵包
King's Bread

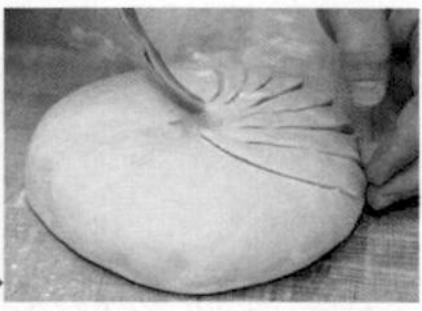

CHEF'S TIPS **To obtain a clean cut, ensure your loaf is not over proofed when cutting and baking. Once egg washed, allow the crust to dry a little before cutting in order to have clean cuts and avoiding wet dough sticking to the blade.**

麵包師建議 **如要呈現清晰的切割口，在進行切割和烘焙時，確保麵糰沒有過度發酵。技巧是於刷蛋液後，讓麵糰外層略乾燥，因過濕的麵糰會黏貼住切割刀。**

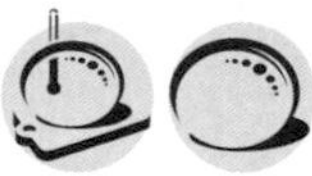

63°C | A: 8mins(S) B: 10mins(F) | 25°C | 45 mins | 450 gm | 15 mins | round | 1 hr | 200°C | 185°C | 55 mins

	INGREDIENTS	RATIO
500 gm	Flour type 45	100%
300 ml	Milk	60%
12 gm	Dry yeast	2.4%
40 gm	Sugar	8%
10 gm	Salt	2%
30 gm	Eggs	6%
1 pc	Lemon zest	
90 gm	Butter	18%
35 gm	Almond paste	7%
1 pc	Vanilla bean	
	Egg wash	
	Cocoa nibs	

	材料	比例
500 克	低筋麵粉	100%
300 毫升	鮮奶	60%
12 克	乾酵母	2.4%
40 克	細砂糖	8%
10 克	鹽	2%
30 克	雞蛋	6%
1 個	檸檬皮	
90 克	奶油	18%
35 克	杏仁醬	7%
1 支	香草豆	
	蛋液適量	
	可可碎豆適量	

METHOD

1 Knead the dough with all ingredients except butter and salt.

2 Add the salt and butter 3 minutes before the end of mixing time.

3 Let the dough rest as it is, covered with a plastic film for the bulk fermentation.

4 Weigh the dough and pre-shape in round shape; allow resting.

5 Shape in round loaves and place them on a baking tray lined with baking paper.

6 Proof to the final fermentation.

7 Gently punch a whole in the center of the loaf with a finger to 75% of the depth.

8 Egg wash and cut the loaf from the outside to the inside, all around. Sprinkle cocoa nibs.

9 Bake with steam until golden brown; open the exhaust once coloring starts.

製 法

1 混合所有材料，除奶油和鹽外，搓揉成麵糰。

2 在完成搓揉過程的最後 3 分鐘，加入奶油和鹽。

3 蓋上一張保鮮膜，鬆弛麵糰至體積增大。

4 量重量和初步整形，做成圓形狀，再次鬆弛麵糰。

5 造成圓長形和置放在已墊烘焙紙的烤盤上。

6 進行最後醒麵過程。

7 用手指在麵糰的中央處，輕輕按壓至 75% 深度，呈一小孔。

8 刷上蛋液，並從外至內沿周邊下刀切割，撒上可可碎豆。

9 噴蒸氣烘焙至金黃色，待開始轉顏色，打開排氣活門。

地瓜焦糖核桃麵包

Sweet Potato & Caramelised Chinese Walnut "Cocotte"

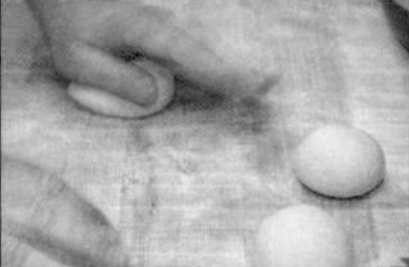

CHEF'S TIPS **Bake your yellow sweet potatoes in the oven will avoid making them watery. Once cold, mash them smoothly before adding to the dough. Depending on the potatoes, you might need to add a little water if the dough is too hard.**

麵包師建議 **用烤爐烘焙地瓜可避免它們有過多水份，待涼後，在加入麵糰前先壓成幼滑薯茸。按照地瓜的狀況，如感覺麵糰有點乾，可加入少許清水調節。**

63°C

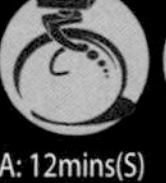

A: 12mins(S)
B: 6mins(F) | 25°C | 45 mins | 50 gm | 10 mins | round rolls | 2.5 hrs | 190°C | 180°C | 35 mins

	INGREDIENTS	RATIO
420 gm	Flour type 45	100%
30gm	Egg yolks	7.14%
50gm	Eggs	11.90%
20 ml	Milk	4.76%
5gm	Dry yeast	1.19%
8gm	Salt	1.90%
75gm	Brown sugar	17.86%
115gm	Butter	27.38%
240gm	Cooked/mashed sweet potatoes	57.14%
1 pcs	Lemon zest	
60gm	Chinese caramelised walnuts, chopped	
	Dusting flour	
	Egg wash	

	材料	比例
420 克	低筋麵粉	100%
30 克	蛋黃	7.14%
50 克	雞蛋	11.90%
20 毫升	鮮奶	4.76%
5 克	乾酵母	1.19%
8 克	鹽	1.90%
75 克	紅糖	17.86%
115 克	奶油	27.38%
240 克	熟地瓜泥	57.14%
1 個	檸檬皮	
60 克	山核桃碎	
	麵粉適量（撒面）	
	蛋液適量	

METHOD

1. Knead the dough; add the salt 3 minutes before kneading ends. Keep walnuts for garnish.
2. Allow resting and weigh the dough; pre-shape it in rolls and rest the dough again.
3. Shape the round rolls and place them in your greased cocotte pan.
4. Proof the buns; apply egg wash and sprinkle with caramelised walnuts.
5. Bake with steam until golden brown; open exhaust once coloration starts.
6. Decorate with optional gold leaves.

製 法

1. 搓揉麵糰，在搓揉過程的最後 3 分鐘加入鹽，續揉至完成。山核桃作裝飾備用。
2. 麵糰鬆弛，量重量。 接着，初步整成卷狀和再次醒麵。
3. 造成圓形，放在已刷油的煎鍋。
4. 將小麵卷醒麵，刷上蛋液和撒上山核桃。
5. 放入蒸氣，烘焙至金黃色。當麵包正開始轉色，打開排氣活門。
6. 可選擇放上金箔作裝飾。

CHEF'S TIPS Simply take a nice dark chocolate and ground it into powder for the filling. Use a good amount of chocolate powder mixture inside the bread to create a nice swirl effect. Pearl sugar will remain white and crisp whereas regular sugar would melt and caramelise.

麵包師建議 把優質純黑巧克力磨成粉狀作餡料。如要製作出漂亮的螺旋效果，應放足夠份量的巧克力粉混合物於麵包裏。用珍珠糖能保持外層白亮和酥脆，如使用正常白糖就會熔掉和變焦糖。

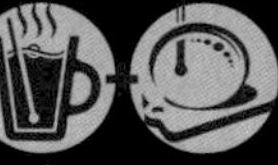

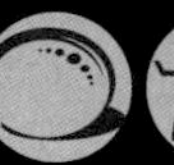

63°C | A: 10mins(S) B: 6mins(F) | 26°C | 1 hr | 350 gm | 10 mins | rolled in mould | 1.5 hrs | 190°C | 190°C | 35 mins

	INGREDIENTS	RATIO
600 gm	Flour type 55	100%
10 gm	Dry yeast	1.67%
30 gm	Milk powder (full fat)	5%
80 gm	White sugar	13.33%
10 gm	Salt	1.67%
60 gm	Egg	10%
250 ml	Fresh milk	41.67%
150 ml	Liquid cream	25%
130 gm	Chocolate 70%	
60 gm	Icing sugar	
50 gm	Pearl sugar	
	Dusting flour	
	Egg wash	

	材料	比例
600 克	中筋麵粉	100%
10 克	乾酵母	1.67%
30 克	奶粉（全脂）	5%
80 克	細砂糖	13.33%
10 克	鹽	1.67%
60 克	雞蛋	10%
250 毫升	鮮奶	41.67%
150 毫升	液態鮮奶油	25%
130 克	70% 可可含量的巧克力	
60 克	糖霜	
50 克	珍珠糖	
	麵粉適量（撒面）	
	蛋液適量	

METHOD

1. Ground your chocolate into powder and mix with icing sugar; reserve for the stuffing.
2. Knead the dough and add the salt 3 minutes before kneading ends.
3. Allow resting; roll the dough in rectangular shape of about 1.5 cm thick and 30 cm in width .
4. Egg wash and spread the cocoa mixture evenly over the dough.
5. Roll the dough as a snail, cut it into chunks to fit your mould in length.
6. Cut your roll in halves and twist the two halves together. Proof.
7. Once proofed, egg wash and sprinkle the pearl sugar.
8. Bake without steam and open exhaust until golden brown.

製 法

1. 把巧克力磨碎，與糖霜混合，作釀餡備用。
2. 搓揉麵糰，待最後搓揉前 3 分鐘，加入鹽。
3. 讓麵糰鬆弛，然後擀碾成長方形1.5厘米厚和30厘米寬。
4. 刷上蛋液，把巧克力混合物均勻地刷在麵糰上。
5. 將麵糰捲成如蝸牛似的螺旋紋，按模具的大小切成塊。
6. 把長卷切半，再把兩半麵糰扭在一起，待醒麵。
7. 當醒麵完成，刷上蛋液和撒上珍珠糖。
8. 不用放蒸氣，直至入烤爐，打開排氣活門，烤至金黃色。

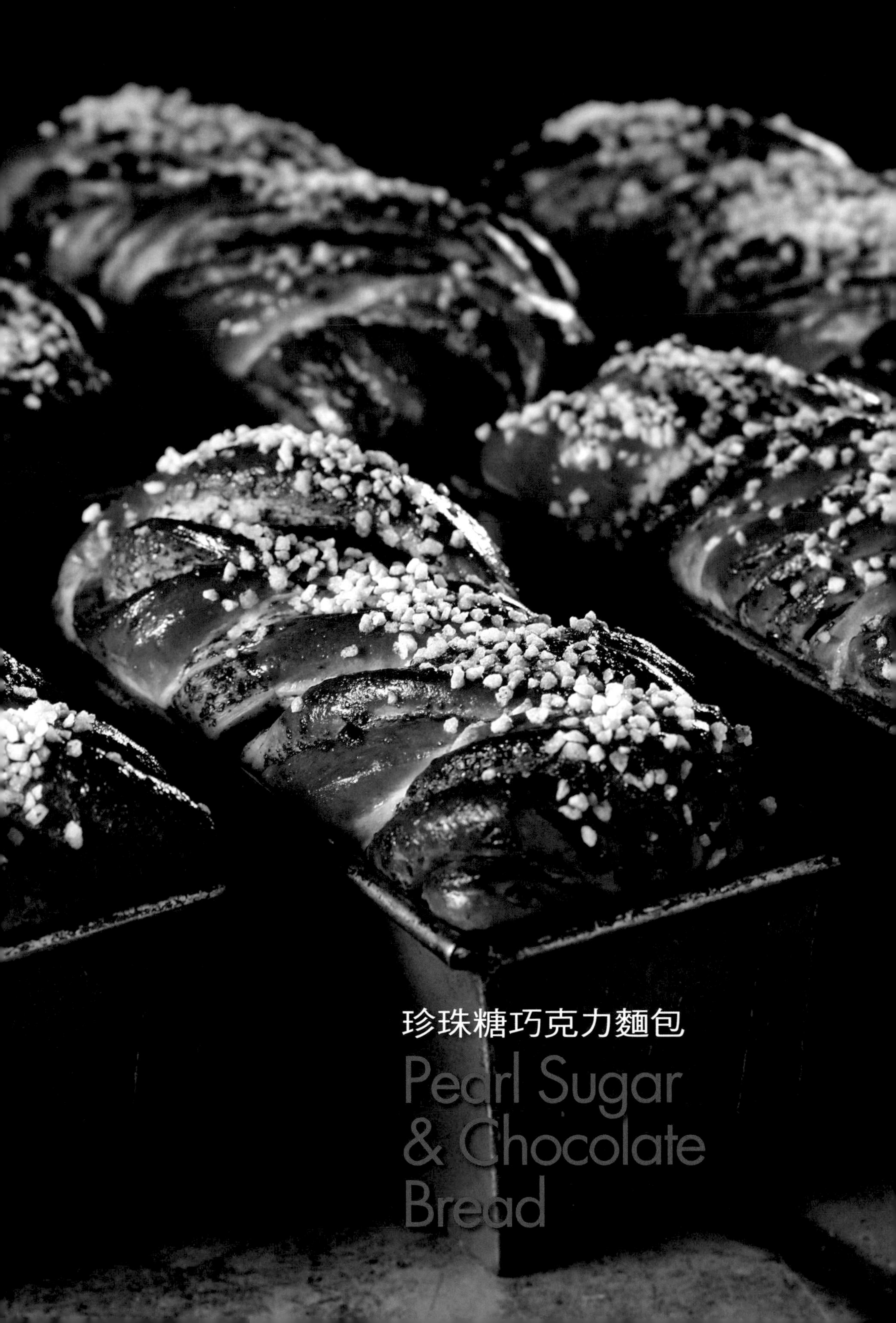

珍珠糖巧克力麵包

Pearl Sugar & Chocolate Bread

榛果巧克力脆卷

Hazelnut Praline Flaky Rolls

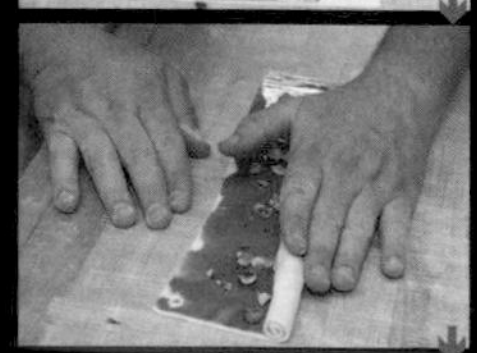

CHEF'S TIPS These flaky hazelnut rolls are as good as the hazelnut praline paste quality you'll use in the making. Try to find pure hazelnut paste, without sugar added to obtain a maximum of flavor. You can finish these rolls with glazing, icing sugar or simply add more roasted hazelnuts on the egg wash, before baking.

麵包師建議 這些榛果脆卷的品質跟榛果醬很貼近。嘗試購買純淨的榛果醬，因沒有添加砂糖能把所有味道帶出。入爐前，在已刷蛋液的榛果麵包卷上，加點亮油、糖霜或只是簡單地多加一些已烘焙的榛果便可。

61˚C | A: 8mins(S) B: 8mins(F) | 24˚C | 30 mins | 3x80 gm | 15 mins | snail rolls | 1 hrs | 200˚C | 190˚C | 24 mins

	INGREDIENTS	RATIO
500 gm	Flour type 45	100%
250 ml	Fresh milk	50%
15 gm	Dry yeast	3%
50 gm	White sugar	10%
10 gm	Salt	2%
25 gm	Eggs	5%
1 pc	Lemon zest	0.2%
200 gm	Butter for folding	40%
100 gm	Hazelnut, roasted	
	Dusting flour	
	Egg wash	
	Filling	
90 gm	Butter	
50 gm	Hazelnut praline paste	
120 gm	Muscovado sugar	

	材料	比例
500 克	低筋麵粉	100%
250 毫升	鮮鮮奶	50%
15 克	乾酵母	3%
50 克	細砂糖	10%
10 克	鹽	2%
25 克	雞蛋	5%
1 個	檸檬皮	0.2%
200 克	奶油（夾心）	40%
100 克	榛果（烘焙）	
	麵粉適量（撒面）	
	蛋液適量（刷面）	
	餡料	
90 克	奶油	
50 克	榛果醬	
120 克	黑糖	

METHOD

1. Prepare the filling; mix the butter and hazelnut praline into a creamy texture, add the brown sugar. Keep aside, at room temperature.
2. Knead the dough and add the salt 3 minutes before kneading ends.
3. After the bulk fermentation, weigh the dough and flatten it into a square. Allow cooling in the fridge for about 3 hours.
4. Roll the folding butter into a sheet of about 1cm in thickness and place it in the middle of your dough. Fold the side towards the center to encase the butter in the dough.
5. Roll the dough gently to 1.5 cm thickness and give it a simple fold.
6. Repeat the step 5 another two times and wrap your dough in plastic film. Keep it in the fridge for 2 hours before using.
7. Roll the dough on a floured surface, to 3 mm in thickness. Spread the hazelnut filling and sprinkle with the chopped hazelnuts.
8. Apply egg wash once on one edge of the dough and roll into snail. Cut and place them in row of 3 pieces stuck together, on a baking tray lined with baking paper.
9. Once fully proofed, apply egg wash, sprinkle chopped hazelnut and bake until golden brown with steam and closed exhaust.

製 法

1. 預備餡料，把奶油和榛果醬混合至滑潤，加入紅糖，置室溫下備用。
2. 搓揉麵糰，待最後搓揉前 3 分鐘，加入鹽。
3. 待大量發酵後，量重量和碾擀至成正方形，置冰箱內約 3 小時。
4. 把奶油夾心碾擀為 1 厘米厚的片裝，放在麵糰中央的位置。將周邊麵糰向內摺覆，包裹奶油。
5. 再把麵糰徐徐碾成 1.5 厘米厚，簡單摺疊。
6. 重複（步驟 5）2 次，用保鮮膜包好麵糰，使用前放進冰箱待 2 小時。
7. 在工作枱上撒粉，碾擀麵糰約 3 毫米厚，抹上榛果醬和撒上榛果碎。
8. 在麵糰的角位刷上一次蛋液，捲成螺旋紋。把麵糰切割，將 3 個小麵糰貼合成一排，放在已鋪紙的烤盤上。
9. 完全醒麵後，刷點蛋液，撒上榛果碎，放入烤爐放蒸氣和關閉排氣活門，烘至金黃色。

葡萄乾胡桃麵包

Pecan and Raisin Boule

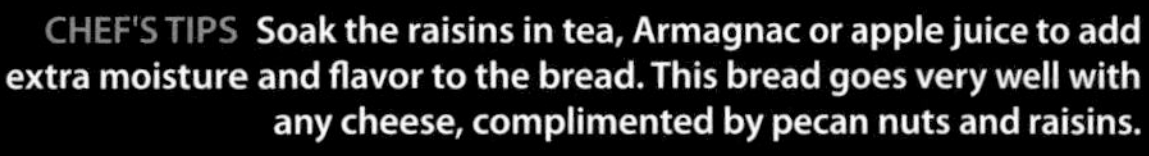

CHEF'S TIPS Soak the raisins in tea, Armagnac or apple juice to add extra moisture and flavor to the bread. This bread goes very well with any cheese, complimented by pecan nuts and raisins.

麵包師建議 可把葡萄乾浸在茶、法國阿瑪涅克白蘭地酒或蘋果汁內，如希望麵包能增添味道和濕潤。這款可配任何起司、美洲胡桃和葡萄乾伴吃，美味可口。

63°C | A: 10mins(S) B: 6mins(F) | 27°C | 1 hr | 450 gm | 15 mins | round | 3 hrs | 225°C | 200°C | 35 mins

	INGREDIENTS	RATIO
650 gm	Flour type 65	100%
75 gm	Whole wheat flour	11.54%
40 gm	Rye flour	6.15%
10 gm	Dry yeast	1.31%
30 gm	Wheat germs	3.92%
20 gm	Salt	2.61%
580 ml	Water	75.82%
250 gm	Raisins	32.68%
150 gm	Pecan nuts	19.61%
	Dusting flour	

	材料	比例
650 克	高筋麵粉	100%
75 克	全麥麵粉	11.54%
40 克	黑麥	6.15%
10 克	乾酵母	1.31%
30 克	小麥胚芽	3.92%
20 克	鹽	2.61%
580 毫升	清水	75.82%
250 克	葡萄乾	32.68%
150 克	美洲胡桃仁	19.61%
	麵粉適量（撒面）	

METHOD

1. Knead all the ingredients except salt, raisins and pecan nuts.
2. Add the salt 3 minutes before the end of mixing time.
3. Thoroughly mix the raisins and pecan nuts at the end of the kneading in slow speed.
4. After the bulk fermentation, weigh the dough and allow resting.
5. Shape the boule and place them in floured proofing wooden basket.
6. Once proofed, turn the baskets on a floured board to slide your loaf directly on the baking stone.
7. Bake with steam and open the exhaust once the bread starts coloring. Bake until dark golden brown.

製 法

1. 把所有材料除鹽、葡萄乾和美洲胡桃外混合，搓揉成麵糰。
2. 在搓揉過程的最後 3 分鐘，加鹽後繼續搓揉完成。
3. 最後，改以慢速隨意地混合葡萄乾和美洲胡桃仁。
4. 初步發酵麵糰後，量重量和讓其鬆弛。
5. 把麵糰整型，放在已撒麵粉的木製籃繼續發酵。
6. 當醒麵完成，把木籃翻在已撒粉的板上，再直接滑落於烘焙石上。
7. 放入蒸氣後烘焙，當麵糰剛開始轉色，立即打開排氣活門，烘至深金黃色。

CHEF'S TIPS **Due to the high coconut content in the dough, you might see traces of coconut oil on the dough, it's normal. Use cold butter to make your crumble and you'll obtain the perfect texture. I used Masarang palm sugar from Indonesia; it gives a unique flavor and intense contrast to the crumble. Moreover, it undertone perfectly the mellow aromas of coconut.**

麵包師建議 **麵糰含的椰子成份含量高，如有椰油痕跡在麵糰是正常的。應用冷凍奶油以獲得完美質感。在此食譜處用了 Masarang 的印尼椰糖，它能帶出獨特香味、跟碎粒形成強烈對比。再者，它與香醇的椰子香氣，配合得天衣無縫。**

63˚C · A: 12mins(S) B: 5mins(F) · 26˚C · 45mins · 50 gm · 10 mins · buns · 1.5 hrs · 200˚C · 190˚C · 25 mins

	INGREDIENTS	RATIO
420 gm	Flour type 45	100%
8 gm	Dry yeast	1.9%
40 gm	Coconut milk powder	9.52%
25 gm	Dessicated coconut	5.95%
25 gm	White sugar	5.95%
12 gm	Salt	2.86%
60 gm	Butter	14.29%
280 ml	Coconut milk	66.67%
	Egg wash	
	Dusting flour	
	Crumble	
50	Butter	
70	Palm sugar	
65	Self raising flour	
25	Dessicated coconut	

	材料	比例
420 克	低筋麵粉	100%
8 克	乾酵母	1.9%
40 克	椰子粉	9.52%
25 克	椰絲	5.95%
25 克	細砂糖	5.95%
12 克	鹽	2.86%
60 克	奶油	14.29%
280 毫升	椰汁	66.67%
	蛋液適量（刷面）	
	麵粉適量（撒面）	
	碎粒	
50 克	奶油	
70 克	棕櫚糖	
65 克	自發麵粉	
25 克	椰絲	

METHOD

1. Prepare the crumble by mixing all the ingredients together and pass it through a grid to make crumbles. Reserve in the fridge.
2. Knead the dough and add the salt 3 minutes before kneading ends.
3. After the bulk fermentation, weigh the dough and allow resting.
4. Shape the bun and place them on a baking tray lined with baking paper.
5. Once fully proofed, apply egg wash and sprinkle crumbles on top of the buns.
6. Bake until golden brown without steam and with open exhaust.
7. Dust icing sugar once cooled.

製 法

1. 碎粒材料混合一起，壓入格篩，做成碎粒，置冰箱備用。
2. 搓揉麵糰，在搓揉過程的最後 3 分鐘，才加入鹽，續至搓揉完成。
3. 初步發酵麵糰完成後，量重量，讓其鬆弛。
4. 整型，放在已鋪烘焙紙的烤盤上。
5. 當完成醒麵，刷上蛋液，撒上碎粒於麵糰面。
6. 不用放蒸氣，打開排氣活門直接烘焙至金黃色。
7. 出爐後待涼，撒上糖霜。

椰糖奶油卷
Coconut Palm Sugar Roll

CHEF'S TIPS **At breakfast, I found this bread served warm and toasted with some salty butter to be the perfect start of the day! Use dark chocolate of at least 60% cocoa content or more to snatch that chocolate bite up!**

麵包師建議 **在早餐時，把這麵包溫熱和烘焙成吐司，再加上鹹味的奶油享用，將會是一天的完美開始。使用可可油含量最少 60% 的黑巧克力或多一點巧克力。**

	INGREDIENTS	RATIO
700 gm	Flour type 65	100%
125 gm	Rye flour	17.86%
70 gm	Cocoa powder	8.48%
60 gm	Butter	7.27%
12 gm	Dry yeast	1.45%
28 gm	Salt	3.39%
620 ml	Water	75.15%
190 gm	Dark chocolate chunks	
140 gm	Frozen sour cherries	

	材料	比例
700 克	高筋麵粉	100%
125 克	黑麥	17.86%
70 克	巧克力粉	8.48%
60 克	奶油	7.27%
12 克	乾酵母	1.45%
28 克	鹽	3.39%
620 毫升	清水	75.15%
190 克	黑巧克力塊	
140 克	冷藏酸櫻桃	

METHOD

1. Knead the dough with all the ingredients except the salt, sour cherries and chopped chocolate.
2. Add the salt 3 minutes before the end of kneading; add the sour cherries and chocolate in slow speed, at the end of kneading.
3. Allow resting, pre-shape round loaves and let the dough rest.
4. Shape round loaves and place them on a floured cloth.
5. Once fully proofed, make sure they don’t stick to the cloth and give an X cut on the top.
6. Bake with steam; open the exhaust after about 20 minutes of baking.

製 法

1. 把所有材料除鹽、酸櫻桃和巧克力碎外混合，搓揉成糰 。
2. 在搓揉過程的最後 3 分鐘，才加入鹽混合，續至完成搓揉，加入酸櫻桃和巧克力以慢速混合。
3. 待其鬆弛，預先整型成圓橄欖球狀，續讓它鬆弛。
4. 造成圓橄欖形，放置在已撒麵粉的發酵布上。
5. 醒麵一旦完成，確認它不黏貼在麵包布，並在包上面畫 “X” 。
6. 放入蒸氣，待烘焙 20 分鐘後打開排氣活門至完成。

黑巧克力酸櫻桃麵包

Dark Chocolate and Sour Cherry Bread

乾果麵包
Thick Dried Fruit Loaf

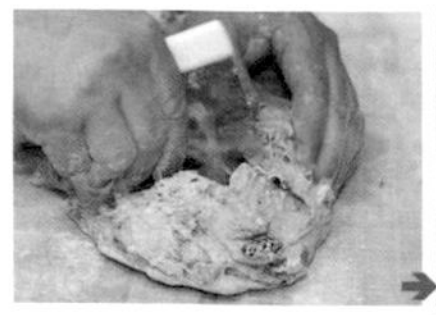

CHEF'S TIPS **The dough will take a longer fermentation time due to its heavy content of fruits and nuts. The loaf will be gaining color faster due to the high content of natural sugar brought in the dough from the dry fruits.**

麵包師建議 **麵糰含高份量的乾果和果仁，需要較長的發酵時間。乾果更帶出高含量的天然糖分，令麵包很快就變色。**

63˚C | A: 15mins(S) B: 2mins(F) | 27˚C | 1 hr | 5x120 gm | 15 mins | round | 2 hrs | 200˚C | 195˚C | 45 mins

	INGREDIENTS	RATIO
450 gm	Flour type 65	100%
120 gm	Rye flour	26.67%
35 gm	Sesame seeds	6.14%
40 gm	Wheat germs	7.02%
12 gm	Salt	2.11%
40 gm	Milk powder	7.02%
12 gm	Dry yeast	2.11%
400 ml	Water	70.18%
	Dusting flour	
	Fruit mix	
30 gm	Dried apricots	
30 gm	Dried prunes	
30 gm	Dried figs	
30 gm	Whole hazelnuts	
30 gm	Walnut halves	
30 gm	Whole almonds	

	材料	比例
450 克	高筋麵粉	100%
120 克	粿麥麵粉	26.67%
35 克	芝麻籽	6.14%
40 克	小麥胚芽	7.02%
12 克	鹽	2.11%
40 克	奶粉	7.02%
12 克	乾酵母	2.11%
400 毫升	清水	70.18%
	麵粉適量（撒面）	
	混合乾果	
30 克	杏脯乾	
30 克	加州梅乾	
30 克	無花果乾	
30 克	整粒榛果	
30 克	半顆美洲胡桃仁	
30 克	整粒杏仁	

METHOD

1. Cut all the dry fruits in large chunks and leave the nuts whole.
2. Knead the dough with all ingredients except the salt and the fruit mix.
3. Add the salt 3 minutes before the end of mixing time. Once the dough is fully kneaded, add the fruits in slow speed.
4. After the bulk fermentation, weight the dough and allow resting.
5. Carefully shape the rounds and place them in a floured long wooden proofing basket.
6. Once fully proofed, flip the bread on a floured board, score each round once and slide the loaf on the baking stone.
7. Bake with steam; open the exhaust once coloring starts.

製 法

1. 所有乾果切成大粒，果仁則保持原粒。
2. 除了鹽和混合乾果外，混合所有材料，搓揉成麵糰。
3. 在搓揉過程的最後 3 分鐘，加入鹽續搓揉完成，加入混合乾果以慢速融合。
4. 初步發酵麵糰完成後，量重量和讓其鬆弛。
5. 小心整成圓形，置放在已撒麵粉的長木質發酵籃。
6. 一旦醒麵完成，翻麵糰置於已撒粉的木板上，並在每個麵糰刻紋，再滑落於烘焙石上。
7. 放入蒸氣後烘焙，當麵糰開始變色，打開排氣活門至完成。

CHEF'S TIPS **With a fairly high amount of eggs, sugar and fat, it is important to watch the oven temperatures near the end of baking as the loaves will gain quick coloring. Also ensure to bake the Cuchaulle in a dry oven, without before steam. You might want to keep your oven's door ajar near the end of baking.**

麵包師建議 **由於麵糰含有較高量雞蛋、砂糖和脂肪，接近最後烘焙的階段，必須小心掌控爐溫，因其很快會上色。同時，烘焙前，確保爐內乾爽沒有預先放入蒸氣，你可以在接近完成烘烤時，把爐門半開。**

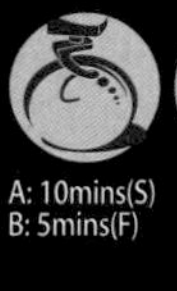

63°C | A: 10mins(S) B: 5mins(F) | 26°C | 30 mins | 350 gm | 15 mins | spiral | 45 mins | 180°C | 180°C | 35 mins

	INGREDIENTS	RATIO
480 gm	Flour type 55	100%
100 gm	Eggs	20.83%
100 ml	Water	20.83%
60 gm	Butter	12.5%
50 gm	White sugar	10.42%
8 gm	Dry yeast	1.67%
10 gm	Salt	2.08%
15 pcs	Saffron pistil	
1 pc	Vanilla bean	
	Icing sugar for dusting	
	White sugar for dusting	
	Water	
	Dusting flour	

	材料	比例
480 克	中筋麵粉	100%
100 克	雞蛋	20.83%
100 毫升	清水	20.83%
60 克	奶油	12.5%
50 克	細砂糖	10.42%
8 克	乾酵母	1.67%
10 克	鹽	2.08%
15 條	番紅花雌蕊	
1 支	香草豆	
	糖霜適量（撒面）	
	細砂糖適量（撒面）	
	清水適量	
	麵粉適量（撒面）	

METHOD

1. Cut the vanilla bean lengthwise and scrap the vanilla seeds in the dough.
2. Knead the dough with all the ingredients and add the salt 3 minutes before kneading ends.
3. After the resting time, shape the dough like baguette and roll it as a snail.
4. Place on baking tray with silicon paper.
5. Once fully proofed, brush the loaves with water and sprinkle completely with white sugar.
6. Dust heavily with icing sugar. Make sure to use regular icing sugar without starch added.
7. Bake without steam and open exhaust until golden brown.

製 法

1. 香草豆剖開，挖出籽。
2. 混合所有材料搓揉成糰，在搓揉過程的最後 3 分鐘，續搓揉至完成。
3. 鬆弛時間完成，整型成棍狀，再捲成螺旋狀。
4. 放入已鋪有矽膠墊的烤盤。
5. 醒麵完成，在麵糰刷點清水，並撒滿細砂糖。
6. 撒上大量糖霜，確保使用沒有添加粟粉的糖霜。
7. 不用放蒸氣於爐內，但需打開排氣活門，烘烤至金黃色便可。

番紅花香草麵包

Saffron Vanilla Cuchaulle

Prune Walnut Bread

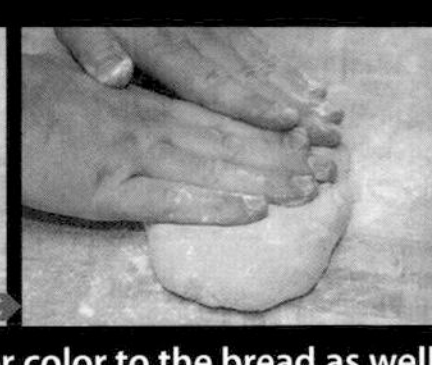

CHEF'S TIPS Prunes will bring a natural darker color to the bread as well as some extra sugar, giving the loaves an extra darker color after baking. Simply cut the prunes in halves will give you larger chunks of fruit in each slice and as well, it will allow you to keep any potential prune pits out of the bread!

The fermented dough is a piece of dough you have left from the day before and stored in the fridge. If you don't have it on hand, you can produce a small dough. To produce the fermented dough, you can use 150 gm of flour type 65 and 100 ml of water with 2 or 3 grams of yeast, knead it, let it ferment for 2 hours at room temperature and leave it in the fridge until the next day before using.

麵包師建議 由於加州梅會帶給麵包天然深色澤和更多糖份，所以烘焙後，它比一般麵包深色。加州梅開半，便能增加加州梅塊的份量，把麵包切開時，可有更多機會見到它們。

已發酵的麵糰是指它早在一日前置於室溫發酵，然後存放於冰箱。如果你沒有預先製作這麵糰，可以製造一小塊，方法非常簡單，只要用上 150 克 65 號的麵粉，100 毫升清水，並與 2 至 3 克乾酵母，搓揉成糰，置室溫下發酵 2 小時，放冰箱貯存，直至翌日應用。

60°C | A: 12mins(S) B: 6mins(F) | 24°C | 1.5 hr | 450 gm | 15 mins | flat | 1.5 hrs | 200°C | 190°C | 45 mins

	INGREDIENTS	RATIO
250 gm	Flour type 55	100%
250 gm	Rye flour type 770	100%
30 gm	Liquid malt	6%
5 gm	Dry yeast	1%
275 gm	Fermented dough	55%
450 ml	Water	90%
17 gm	Salt	3.4%
310 gm	Dried prune	62%
160 gm	Walnut halves	32%

	材料	比例
250 克	中筋麵粉	100%
250 克	770 號粿麥麵粉	100%
30 克	流質麥芽糖	6%
5 克	乾酵母	1%
275 克	發酵麵糰	55%
450 毫升	清水	90%
17 克	鹽	3.4%
310 克	加州梅乾	62%
160 克	半顆美洲胡桃	32%

METHOD

1. Knead the dough with all the ingredients except the walnuts and the prune.
2. Add the salt 3 minutes before kneading ends.
3. After the resting time, pre-shape the dough in flat slabs.
4. Shape the breads in flat rectangles and place on a floured cloth.
5. Once fully proofed, slide the loaf directly on baking stone.
6. Bake until golden brown with steam; open exhaust once coloring starts.

製 法

1. 所有材料除美洲胡桃和加州梅外混合，搓揉成麵糰。
2. 在揉麵過程的最後 3 分鐘，加入鹽續搓揉至完成。
3. 鬆弛麵糰的時間到了，把麵糰整成平厚板狀。
4. 再整成長平方形，放在已撒粉的布上。
5. 一旦完成後，直接滑落在烘焙石上。
6. 放入蒸氣，烘焙至金黃色，當轉色時立即打開排氣活門。

洋梨杏仁塔

Pear Frangipane Brioche

63˚C | A: 12mins(S) B: 5mins(F) | 25˚C | 1 hr | 90 gm | 10 mins | flat round | 30 mins | 200˚C | 190˚C | 25 mins

	INGREDIENTS	RATIO
	Brioche	
500 gm	Flour type 45	100%
12 gm	Sugar	2.4%
130 gm	Butter	26%
180 gm	Egg Yolks	36%
12 gm	Salt	2.4%
80 ml	Fresh Milk	16%
10 gm	Dry Yeast	2%
12 gm	Malt	2.4%
	Egg wash	
	Dusting flour	
	Frangipane	
150 gm	Butter	
150 gm	White sugar	
150 gm	Eggs	
150 gm	Almond powder	
40 gm	Self raising flour	
190 gm	Pastry cream	
	A pinch of salt	
	Poached pears	

	材料	比例
	奶油塔	
500 克	低筋麵粉	100%
12 克	砂糖	2.4%
130 克	奶油	26%
180 克	蛋黃	36%
12 克	鹽	2.4%
80 毫升	鮮奶	16%
10 克	乾酵母	2%
12 克	麥芽	2.4%
	蛋液適量（刷面）	
	麵粉適量（撒面）	
	杏仁奶油餡	
150 克	奶油	
150 克	細砂糖	
150 克	雞蛋	
150 克	杏仁粉	
40 克	自發粉	
190 克	卡士達醬	
	鹽一小撮	
	水煮梨適量	

METHOD

1. Knead the dough and add the salt 3 minutes before kneading ends.
2. Let the dough rest as it is, covered with a plastic film for the bulk fermentation.
3. Weigh the dough and pre-shape in round shape; allow resting.
4. Roll the tart on a floured table at around 1cm thick and cut a disc of dough of about 20 cm.
5. Place them on a baking tray lined with baking paper. Allow to proof to ¾.
6. Press gently the middle part of your dough, leaving a 2cm rim to be egg washed.
7. Spread some frangipane mixture on the center of the tart.
8. Add your sliced pears on the frangipane mixture.
9. Bake without steam and with open exhaust until golden brown.
10. Once baked, brush the pear lightly with pear or apricot marmalade, sprinkle with icing sugar.

製 法

1. 搓揉麵糰，在搓揉過程的最後 3 分鐘，加入鹽續搓揉完成。
2. 蓋上一保鮮膜，讓麵糰鬆弛。
3. 量重量，預先做出圓形狀，讓麵糰鬆弛。
4. 在已撒麵粉的檯子上，把奶油塔麵糰碾擀至約 1 厘米厚，鋑出一個直徑約 20 厘米的圓形。
5. 放在已鋪烘焙紙的烤盤上，讓其發酵至 3/4。
6. 在麵糰中央處輕輕壓下，除邊緣的 2 厘米外，刷上蛋液。
7. 在塔的中央，抹上杏仁奶油餡。
8. 加入切成片的水煮梨於杏仁奶油餡。
9. 不用放入蒸氣於烤爐，但需開啟排氣活門，烘焙至金黃色。
10. 烘焙完成，立即徐徐塗上洋梨或杏桃果醬，再撒上糖霜。

CHEF'S TIPS **Poach your pears in pear juice with some cinnamon, cloves and star anise to add extra flavors to your tart. Brush a little pear and vanilla jam on the fruits to give it an extra shine.**

麵包師建議 **把洋梨浸於有肉桂、丁香和八角的洋梨汁，添加洋梨杏仁塔的風味。再刷上少許洋梨和香草果醬，能為水果表面增加額外光澤。**

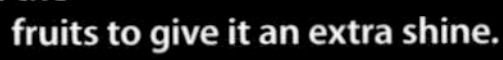
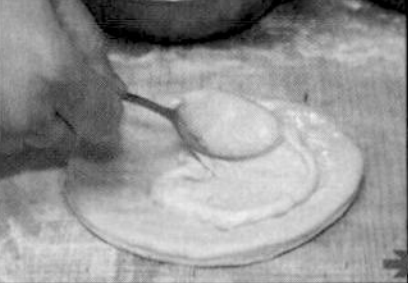

The arrogance of Salt

了不起的鹹麵包

大蘑菇鄉村麵包

Portobello Mushroom Country Bread

65˚C | A: 12mins(S) B: 8mins(F) | 26˚C | 1 hr | 450 gm | 15 mins | round loaf | 45 mins | 235˚C | 210˚C | 45 mins

	INGREDIENTS	RATIO
650 gm	Flour type 65	100%
100 gm	Whole wheat flour	15.38%
100 gm	Rye flour	15.38%
12 gm	Dry yeast	1.41%
35 gm	Wheat germ	4.12%
20 gm	Salt	2.35%
550 ml	Water	64.71%
300 gm	Portobello mushroom	35.29%
2 gm	Cracked black pepper	0.24%
	Dusting flour	

	材料	比例
650 克	高筋麵粉	100%
100 克	全麥麵粉	15.38%
100 克	粿麥麵粉	15.38%
12 克	乾酵母	1.41%
35 克	小麥胚芽	4.12%
20 克	鹽	2.35%
550 毫升	清水	64.71%
300 克	義大利大蘑菇	35.29%
2 克	粗粒黑胡椒	0.24%
	麵粉適量（撒面）	

METHOD

1. Wash and cut the fresh mushrooms into thick slices. Cook them with olive oil and salt.
2. Knead the dough; add salt 3 minutes before kneading ends.
3. Once the kneading is done, add the cooled mushrooms in slow speed.
4. Keep a part of dough in the fridge to later roll the top part of the loaves.
5. After the bulk fermentation, place the dough on a table dusted with fine semolina and weigh the dough. Gently pre-shape round loaves and allow resting.
6. Roll the chilled dough at 3mm thickness and cut discs of the same diameter as your loaves.
7. Place the discs on a cloth dusted with flour and shape your round loaves.
8. Brush water on the top of the loaves and place them upside-down on the discs.
9. Once fully proofed, flip the breads on a floured board.
10. Bake directly on your baking stone with steam. Open the exhaust once coloring starts.

製法

1. 把新鮮大蘑菇清洗和切厚片，以橄欖油和鹽炒軟。
2. 搓揉麵糰，在搓揉過程的最後 3 分鐘，加入鹽續搓揉完成。
3. 揉好麵後，加入放涼的蘑菇以慢速拌勻。
4. 把部份麵糰置冰箱，最後擀薄平放在麵包上面。
5. 基本發酵麵糰完成後，放在已撒粗粒小麥粉的工作檯上，量重量。輕輕地做麵糰的初整型，整成圓形狀，讓其鬆弛。
6. 擀平麵糰，厚約 3 毫米，把圓紙片切割成相同直徑的圓麵包。
7. 放在已撒麵粉的布上，做成一個個圓麵包。
8. 在麵包上刷一層清水，將麵糰反轉放在大碗裡面。
9. 當發酵完成，將麵糰翻轉放在已撒麵粉的木板上。
10. 放入蒸氣，直接放入烘焙石烘烤。當麵包開始轉色，開啟排氣活門。

CHEF'S TIPS You can use any type of mushrooms; however, grilled Portobello mushrooms have a specific flavor that mixes well in bread, especially char-grilled. Mushroom bread is great toasted and served warm with any type of soups or melted cheese sandwiches and tartines. Naturally, mushrooms contain a lot of water, therefore the amount of water is reduced in the recipe or else the dough will be too soft.

麵包師建議 此食譜用了義大利大蘑菇，烘焙方法可把其特別香味帶出，配上麵包可口無比，尤其用炭燒方法烹煮，用其他菇類代替也可。蘑菇麵包最佳的吃法是烘烤，然後放上起司做三明治或法式或意式三文治，配以各種湯類，非常美味。

番茄乾義式培根卷

Sun Dried Tomatoes & Pancetta Rolls

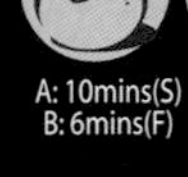

63°C | A: 10mins(S) B: 6mins(F) | 26°C | 45 mins | 80 gm | 15 mins | rolls | 45 mins | 200°C | 190°C | 35 mins

	INGREDIENTS	RATIO
380 gm	Flour type 65	100%
150 ml	Milk	39.47%
150 ml	Water	39.47%
40 ml	Olive oil	10.53%
15 gm	Salt	3.95%
10 gm	Dry yeast	2.63%
80 gm	Sun Dried Tomato	21.05%
SQ	Tomato sauce	
SQ	Pancetta	
SQ	Fresh basil	
	Dusting flour	
	Egg wash	
	Dried herbs	

	材料	比例
380 克	高筋麵粉	100%
150 毫升	鮮奶	39.47%
150 毫升	清水	39.47%
40 毫升	橄欖油	10.53%
15 克	鹽	3.95%
10 克	乾酵母	2.63%
80 克	風乾番茄	21.05%
	番茄醬適量	
	義大利培根適量	
	新鮮羅勒適量	
	麵粉適量（撒面）	
	蛋液適量（刷面）	
	乾燥香草適量	

METHOD

1 Cook your tomato sauce with salt and black pepper and reduce it by one third to obtain a thicker tomato sauce.

2 Knead the dough and add the salt 3 minutes before kneading ends. Add the chopped sun dried tomatoes at the end of kneading in slow speed.

3 After resting time, scale the dough and pre-shape in round.

4 Once rested, roll the dough into discs of about 12cm and spread the tomato sauce, leaving a rim of 1cm around the disc of dough.

5 Add the pancetta in pieces and the roughly chopped basil leaves.

6 Fold the disc into a roll and cut the middle part all the way through. Push the inner part to come out slightly in order to expose the inner layers.

7 Place on a baking tray with baking paper and proof.

8 Once fully proofed, brush with egg wash and garnish with pancetta, cheese and basil.

9 Bake until golden brown with steam; open exhaust once coloring starts.

製 法

1 番茄醬加入鹽和黑胡椒，煮到剩下 1/3 變成濃郁番茄醬。

2 搓揉麵糰，在搓揉過程的最後 3 分鐘，加入鹽續搓揉至完成，以慢速拌入風乾番茄。

3 鬆弛後，秤量麵糰， 預先做圓形。

4 待再次鬆弛完成，把麵糰擀成直徑 12 厘米的圓平麵糰，抹上番茄醬，但周邊 1 厘米處，不用抹醬。

5 加入義大利培根和羅勒粗絲。

6 把圓平麵糰捲成卷狀，並在中央部份切割，微微將內部呈現出來，以突出內部層次。

7 放入已鋪烘焙紙的烤盤上。

8 醒麵完成，刷上蛋液，放上義式培根、起司和羅勒裝飾。

9 放入蒸氣，烘烤至金黃色，待麵包轉色便開啟排氣活門。

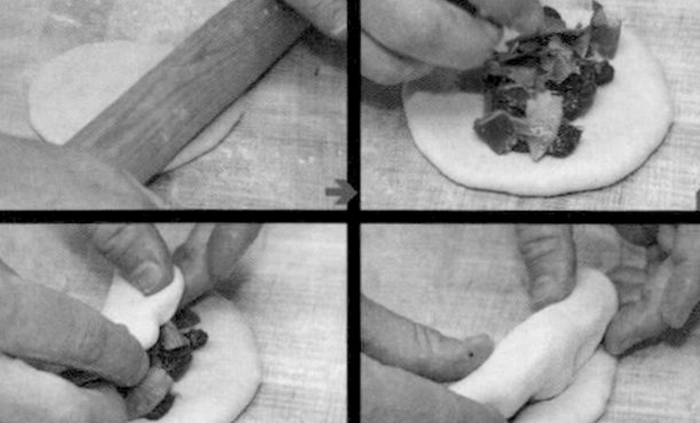

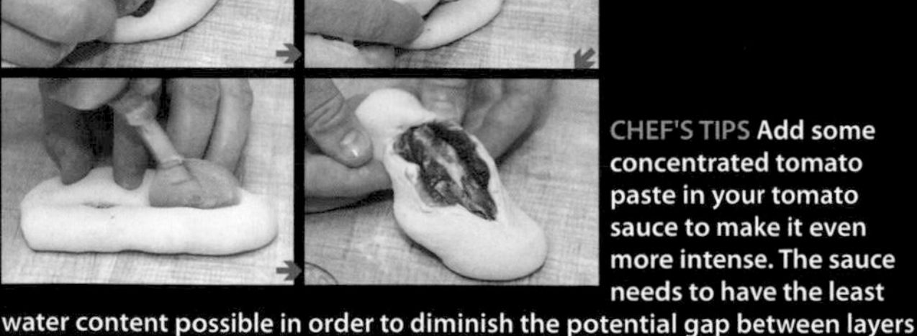

CHEF'S TIPS **Add some concentrated tomato paste in your tomato sauce to make it even more intense. The sauce needs to have the least water content possible in order to diminish the potential gap between layers of bread once baked.**

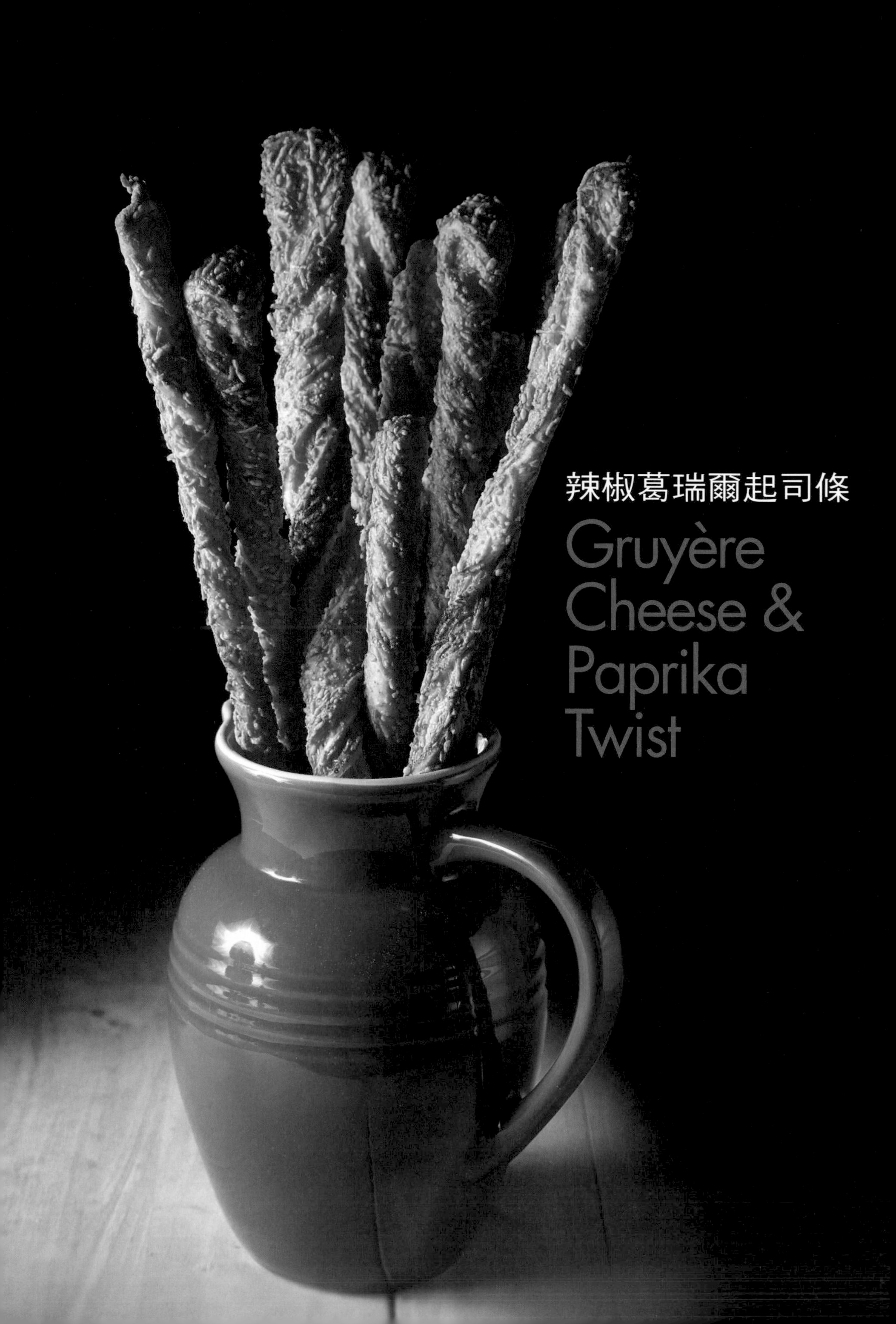

辣椒葛瑞爾起司條

Gruyère Cheese & Paprika Twist

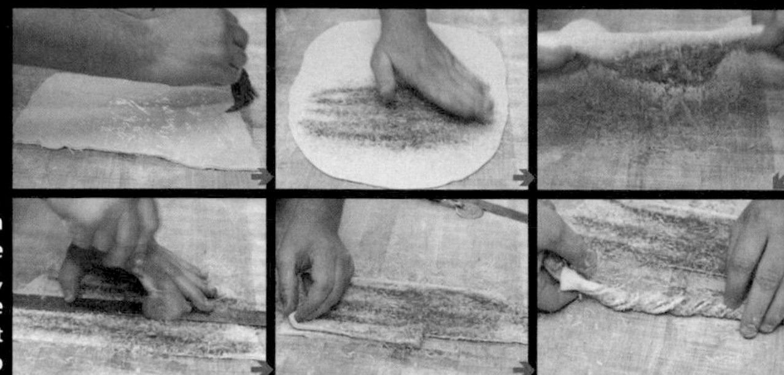

CHEF'S TIPS You can use different cheese to make the twist, such as blue cheese or parmesan, but Gruyere fits the job very well. The olive oil in the dough brings a little extra crunch to the sticks. If they're not crunchy, let them dry a little longer in a cooler oven. We don't let these sticks proofing before baking as we want to obtain that tough and rough look.

麵包師建議 你可以用不同起司做麵包條，例如藍黴起司或義大利帕馬森起司，但葛瑞爾起司的味道較適合。有橄欖油的麵糰會帶給麵包條額外酥脆感。如果它們不夠鬆脆，可把它留在已降溫的烤箱長一點時間才取出，讓它烘乾一點，有較佳效果。如想獲得較堅硬和粗糙的外觀，烘焙前不用讓這些起司條最後發酵。

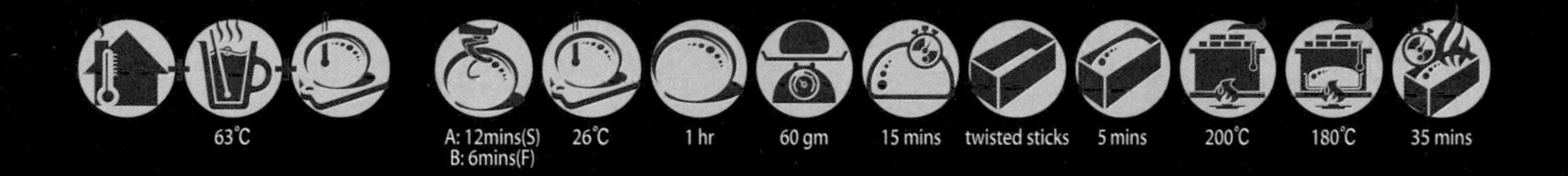

	INGREDIENTS	RATIO
410 gm	Flour type 65	100%
6 gm	Dry yeast	1.46%
310 ml	Water	75.61%
14 gm	Salt	3.41%
12 ml	Olive oil	2.93%
SQ	Gruyere cheese	
SQ	Paprika powder	
	Dusting flour	
	Egg wash	

	材料	比例
410 克	高筋麵粉	100%
6 克	乾酵母	1.46%
310 毫升	清水	75.61%
14 克	鹽	3.41%
12 毫升	橄欖油	2.93%
	葛瑞爾起司適量	
	紅椒粉適量	
	麵粉適量（撒面）	
	蛋液適量（刷面）	

METHOD

1. Knead the dough and add the salt 3 minutes before kneading ends. Keep the cheese and paprika for the finishing. Keep the dough in the fridge for the rest time (1 hour).
2. After resting time, roll the dough between 3 mm to 5 mm thick.
3. Brush egg wash all over the dough and sprinkle heavily with grated gruyere cheese.
4. Sprinkle paprika powder and flip the slab of dough. Repeat the operation on the other side.
5. Using a cutting wheel, cut stripes of 3 cm in width.
6. Fold the stripes by half lengthwise and twist them; place on a baking tray with baking paper and proof.
7. Once rested, bake until golden brown with steam; open exhaust once coloring starts.

製 法

1. 搓揉麵糰，在搓揉過程的最後 3 分鐘，加入鹽續搓揉，完成後拌入起司和紅椒粉。將麵糰置冰箱鬆弛 1 小時。
2. 鬆弛完成，展開麵糰約 3 毫米至 5 毫米厚。
3. 在麵糰上均勻地刷蛋液，撒滿葛瑞爾起司。
4. 用齒輪式切麵刀切成 3 厘米寬的麵皮。
5. 撒上紅椒粉，快速翻轉已切成厚片的麵糰，再撒上紅椒粉。
6. 對摺寬麵皮，扭轉在一起，放入已鋪烤盤紙的烤盤上，進行初步發酵。
7. 一旦麵糰鬆弛完成，在烤箱內放入蒸氣，烘焙至金黃色，待麵包轉色時，打開排氣活門。

法式小洋蔥
Baby Onion and Chervil Bites

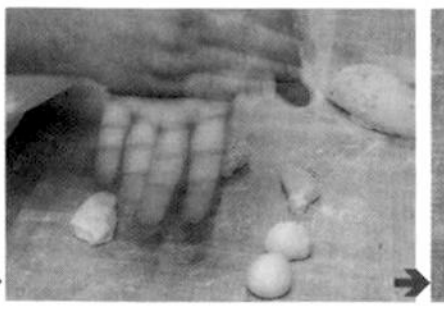

CHEF'S TIPS **The onion confit texture should be melting, and not mushy. It is important to cook it on very slow heat for about 45 minutes to 1 hour. To enhance it, you can add a little orange zest if you like citrus flavor.**

麵包師建議 **紅酒燜洋葱的質地應該是像溶化般的而不是呈柔軟糊狀。方法是必須以慢火烹煮洋葱約 45 分鐘至 1 小時左右。如果你想增添酸果味道，可加點橙皮豐富其食味。**

63℃

A: 10mins(S) B: 5mins(F)
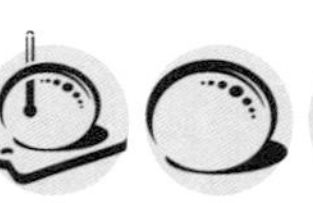
25℃ 2.5 hrs

15 gm 10 mins

small discs

1.5 hrs

220℃

200℃

18 mins

	INGREDIENTS	RATIO
400 gm	Flour type 65	100%
250 ml	Fresh milk	62.5%
60 gm	Butter	15%
20 gm	Sugar	5%
7 gm	Dry yeast	1.75%
80 gm	Eggs	20%
25 gm	Dried onion flakes	6.25%
10 gm	Salt	2.5%
	Baby Chervil	
	Egg wash	
	Onion confit	
3 pcs	Red onion	
2 tbsp	Butter	
2 tbsp	Olive oil	
150 ml	Merlot red wine	
2 pcs	Bay leaves	
1 tsp	Brown sugar	
2 tbsp	Balsamic vinegar	
SQ	Salt and pepper	

	材料	比例
400 克	高筋麵粉	100%
250 毫升	鮮鮮奶	62.5%
60 克	奶油	15%
20 克	砂糖	5%
7 克	乾酵母	1.75%
80 克	雞蛋	20%
25 克	乾洋葱片	6.25%
10 克	鹽	2.5%
	山蘿蔔苗適量	
	蛋液適量（掃面）	
	紅酒燜洋葱	
3 個	紅洋葱	
2 湯匙	奶油	
2 湯匙	橄欖油	
150 毫升	梅洛紅酒	
2 片	月桂葉	
1 茶匙	紅糖	
2 湯匙	義大利陳年葡萄醋	
	鹽和胡椒粉適量	

METHOD

1 Prepare the onion confit; peel and slice the red onions into slices.
2 In a pot, slowly cook the onion with the butter and olive oil and make them sweat.
3 Add the red wine, bay leaves and brown sugar and cook slowly until soft.
4 Finally, add the balsamic vinegar, salt and pepper and adjust the taste.
5 Knead the dough and add the salt 3 minutes before kneading ends.
6 Once rested, weigh the dough and pre-shape into round pieces; allow resting.
7 Shape the round buns and proof completely.
8 Once proofed, brush with egg wash and bake with steam and closed exhaust. Open the exhaust once coloring starts.
9 Once baked, cut them in halves and toast them with a little butter; add a generous amount of confit in the middle and some baby chervil.

製 法

1 紅酒燜洋葱的做法是先把紅洋葱去衣和切片。
2 在鍋內，放點奶油和橄欖油於鍋內慢煮洋葱，令其汁液流出。
3 加入紅酒、月桂葉、紅糖續慢煮至軟。
4 最後，加入義大利陳年葡萄醋、鹽和胡椒，調味。
5 搓揉麵糰，在搓揉過程的最後 3 分鐘，加入鹽續搓揉至完成。
6 一旦鬆弛完成，量重量，把麵糰整出初型，像圓球狀，讓其鬆弛。
7 再把麵糰整成圓形，完成醒麵。
8 醒麵後，刷上蛋液，在烤爐放入蒸氣，然後關閉排氣活門，待烘烤至麵糰轉色，便要開啟排氣活門。
9 烘焙完成，把麵包切半，塗少許奶油烘烤，在其中央加入足夠份量的紅酒燜洋葱和山蘿蔔嬰便成。

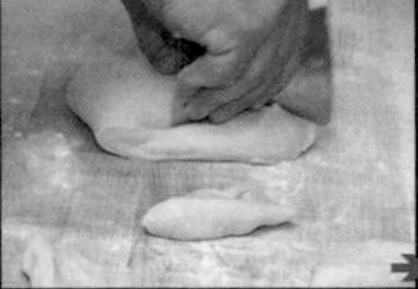
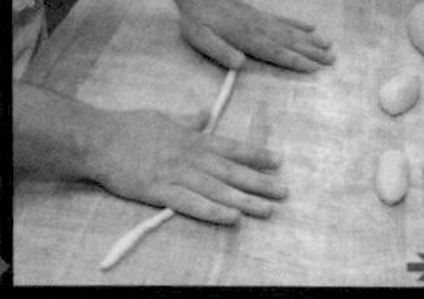

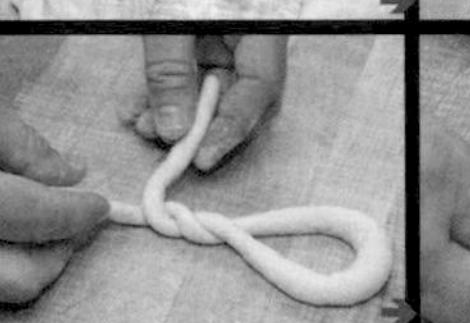
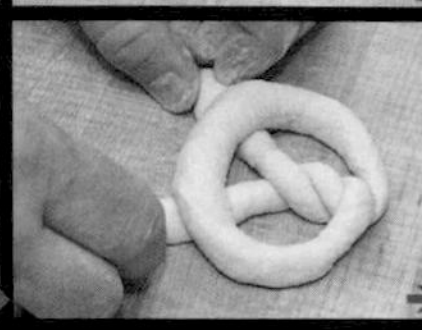

CHEF'S TIPS The dough needs to be harder than other dough and conducted in a faster way in order to obtain chewy and good looking pretzel.

Be sure not to touch the lye with your bare hand and wear gloves and proper protective cloth while working with it. Lye is corrosive and will cause skin injury. If in contact, wash under water with soap right away. Once baked, lye is no longer corrosive and is safe to be consumed.!

麵包師建議 麵糰需要比其他麵糰的質地硬實，以短時間內造型，獲得柔軟而會黏質感和優美脆餅外型。

工作時確保不要徒手觸及鹼水，必須穿上手套和保護衣物。鹼水是腐蝕性物質，會傷害皮膚。如果接觸到，立即用肥皂於清水下沖洗。經烘焙後，鹼水的侵蝕力便失去，可以安全進食。

	63°C		A: 10mins(S) B: 6mins(F)	27°C	1 hr	50 gm	5 mins	pretzel	15 mins	220°C	200°C	25 mins

	INGREDIENTS	RATIO
600 gm	Flour type 65	100%
15 gm	Dry yeast	2.5%
250 ml	Water	41.67%
10 gm	Salt	1.67%
15 ml	Olive oil	2.5%
5 gm	Sugar	0.83%
60 gm	Tomato concentrate	10%
SQ	Dried Chilli Flakes	
SQ	Lye (Caustic soda)	
	Dusting flour	

	材料	比例
600 克	高筋麵粉	100%
15 克	乾酵母	2.5%
250 毫升	清水	41.67%
10 克	鹽	1.67%
15 毫升	橄欖油	2.5%
5 克	砂糖	0.83%
60 克	濃縮番茄	10%
	乾辣椒片適量	
	鹼水適量	
	麵粉適量（撒面）	

METHOD

1. Knead the dough and add the salt 3 minutes before kneading ends.
2. After resting time, shape the bread into pretzel shape and allow them to proof.
3. Prepare your lye bath in a container. Wear protective cloth, gloves and glasses.
4. Deep your pretzel in the lye, turn it over and take it out with a strainer ladle to remove the excess lye.
5. Place on a tray lined with baking paper and sprinkle with dried chili.
6. Bake immediately until dark brown, without steam and with exhaust opened.

製 法

1. 搓揉麵糰，在搓揉過程的最後 3 分鐘，加鹽續搓揉至成。
2. 鬆弛完成後，把麵糰整成椒鹽脆卷餅的螺絲狀，讓其發酵。
3. 準備一盤鹼水，穿上保護性衣著、手套和眼鏡。
4. 把椒鹽脆卷餅浸入鹼水，翻轉，用有柄篩撈出，除去多餘的鹼水。
5. 放入已鋪烘焙紙的烤盤，撒上乾辣椒。
6. 立即放烤爐烘烤至深啡色，不需放蒸氣於爐內，但要開啟排氣活門。

德國番茄辣椒椒鹽脆餅
Tomato Chili Pretzel

大蒜奧勒岡扁麵包
Confit Garlic
& Oregano
Schiacciata

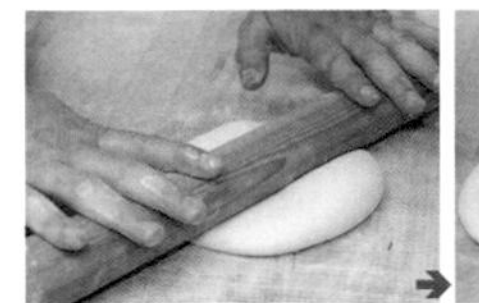  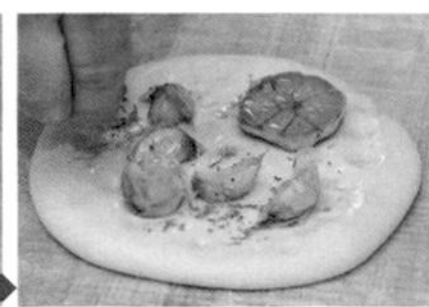

CHEF'S TIPS **To prepare garlic confit, prepare a cooking pot with olive oil and simmer at low temperature the fresh garlic knob until soft and cooked through.**

麵包師建議 **油浸烤蒜頭，先預備一個已放橄欖油的煮鍋，慢火煮鮮蒜頭到軟化和全熟。**

63˚C

A: 10mins(S)
B: 6mins(F)

27˚C

45 mins

350 gm

15 mins

flat bread

30 mins

225˚C

210˚C

40 mins

	INGREDIENTS	RATIO
	Starter	
250 gm	Flour type 45	100%
12 gm	Dry yeast	4.8%
240 ml	Water	96%
	Dough	
400 gm	Flour type 45	100%
100 gm	Eggs	25%
10 gm	Sugar	2.5%
85 ml	Olive oil	21.25
12 gm	Salt	3%
SQ	Fresh oregano	
SQ	Confit Garlic knob	
SQ	Sea salt	
SQ	Olive oil for brushing	
	Dusting flour	

METHOD

1. Mix the starter using a whisk and keep in a warm place for 2 hours.
2. Knead the dough, adding the starter and some oregano in the dough.
3. Add the salt 3 minutes before kneading ends.
4. After resting time, shape the bread into round flat breads.
5. Place on a tray lined with baking paper and proof.
6. Once proofed, brush with olive oil and make holes with your fingers.
7. Push garlic confit knobs in the dough and sprinkle with sea salt and oregano.
8. Bake until golden brown, without steam and with exhaust opened.

	材料	比例
	麵種	
250 克	低筋麵粉	100%
12 克	乾酵母	4.8%
240 毫升	清水	96%
	麵糰	
400 克	低筋麵粉	100%
100 克	雞蛋	25%
10 克	砂糖	2.5%
85 毫升	橄欖油	21.25
12 克	鹽	3%
	新鮮奧勒岡適量	
	油漬烤蒜頭適量	
	海鹽適量	
	橄欖油適量（刷面）	
	麵粉適量（撒面）	

製 法

1. 麵種材料用打蛋器拌勻，在溫暖的地方置放 2 小時。
2. 搓揉麵糰，加入麵種和一些奧勒岡拌勻。
3. 在搓揉過程的最後 3 分鐘，放入鹽續搓揉至完成。
4. 當麵糰鬆弛完成，整型成圓扁狀。
5. 放入已鋪烘焙紙的烤盤上，發酵麵糰。
6. 一旦發酵完成，刷上橄欖油，並用手指在麵糰上挖些小孔洞。
7. 把油漬烤蒜頭壓入麵糰內，再撒上海鹽和奧勒岡。
8. 不用放蒸氣於烤箱內，但需開啟排氣活門，烘烤至金黃色。

希臘黑橄欖法國麵包

Kalamata Olives Baguette

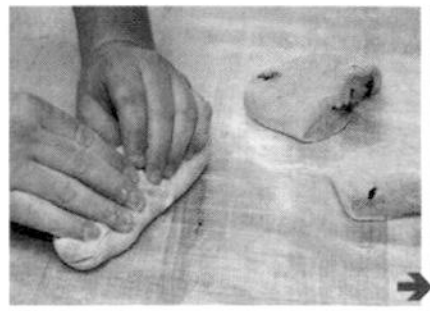
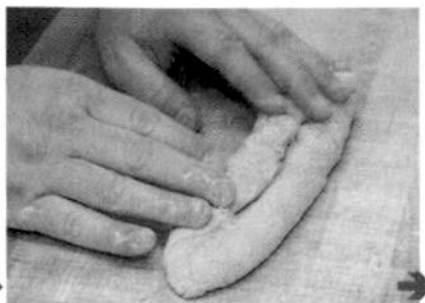
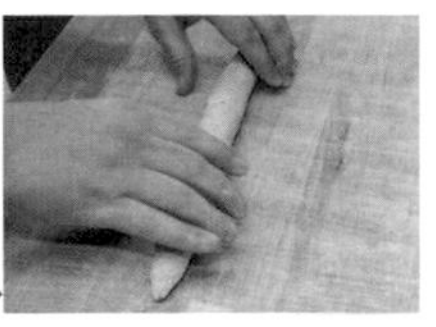

CHEF'S TIPS **The quality of Kalamata olives used is the key to a great flavor in this bread. Using the original Calamata Brand olives, and olive oil for that matter, will help you obtain the finest results and true flavor of Kalamata.**

麵包師建議 **優質 Kalamata 橄欖是帶出特別風味的主要關鍵。選用 Calamata 品牌橄欖和橄欖油，給予你最佳效果和 Kalamata 的真正味道。**

62°C | A: 10mins(S) B: 8mins(F) | 25°C | 1hr | 450 gm | 20 mins | baguette | 45 mins | 225°C | 200°C | 35 mins

	INGREDIENTS	RATIO
300 gm	Flour type 65	100%
50 gm	Whole wheat flour	
50 gm	Rye flour	
6 gm	Dry yeast	2%
15 gm	Wheat germs	5%
12 gm	Salt	4%
325 ml	Water	81.25%
90 gm	Kalamata olives	30%
SQ	Dusting flour	

	材料	比例
300 克	高筋麵粉	100%
50 克	全麥麵粉	
50 克	粿麥麵粉	
6 克	乾酵母	2%
15 克	小麥胚芽	5%
12 克	鹽	4%
325 毫升	清水	81.25%
90 克	Kalamata 橄欖	30%
	麵粉適量（撒面）	

METHOD

1 Knead the dough; add the salt 3 minutes before kneading ends.
2 Once the kneading is done, add the whole olives in slow speed.
3 After resting time, pre-shape in baguette.
4 Allow resting and shape the bread into baguettes, slightly dusted with flour.
5 Proof your breads on a floured cloth.
6 Once proofed, score your bread with a long cut and place it on your baking stone.
7 Bake until golden brown with steam; open the exhaust once coloring starts.

製 法

1 搓揉麵糰，在搓揉過程的最後 3 分鐘，加鹽續搓至完成。
2 一旦搓揉完成，以慢速拌入原粒橄欖。
3 當麵糰鬆弛完成後，初整型，呈長棍狀。
4 鬆弛麵糰後，再次滾成長形，輕輕撒上麵粉。
5 放在發酵布上進行發酵麵包。
6 發酵完成後，沿麵糰長度，放在烘焙石上。
7 放入蒸氣於爐內，烘焙至金黃色，當其開始轉色，開啟排氣活門。

新鮮迷迭香麵包

Fresh Rosemary Pavé

CHEF'S TIPS **The roughly chopped fresh rosemary will release its flavor while baked. Through the process of fermentation, the dough will also impregnate itself with the rosemary essential oil. Using dry rosemary wouldn't achieve the same result.**

麵包師建議 **把新鮮迷迭香略粗切，烘焙時把其味道散發，透過發酵過程，麵糰會同時把迷迭香香精油完全滲入，如用乾迷迭香便不會達到相同效果。**

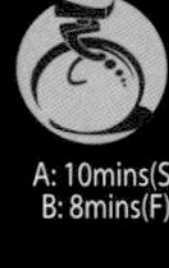
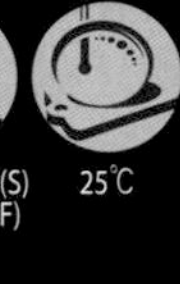
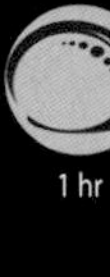

64℃ | A: 10mins(S) B: 8mins(F) | 25℃ | 1 hr | 450 gm | 15 mins | rectangle | 1.5 hrs | 235℃ | 210℃ | 40 mins

	INGREDIENTS	RATIO
400 gm	Flour type 65	100.00
6 gm	Dry yeast	1.5%
15 gm	Salt	3.75%
15 gm	Fresh rosemary	3.75%
300 ml	Water	75%
20 ml	Olive oil	5%
	Fine semolina	
	Fresh rosemary sprigs	
	Sea Salt	

	材料	比例
400 克	高筋麵粉	100.00
6 克	乾酵母	1.5%
15 克	鹽	3.75%
15 克	新鮮迷迭香	3.75%
300 毫升	清水	75%
20 毫升	橄欖油	5%
	粗粒小麥粉	
	新鮮迷迭香條適量	
	海鹽適量	

METHOD

1. CChop the rosemary into rough pieces.
2. Knead the dough; add the chopped rosemary and salt 3 minutes before kneading ends.
3. After the bulk fermentation, place the dough on a table dusted with fine semolina and weight the pieces in natural flatten shape.
4. Place on a cloth dusted with fine semolina to proof.
5. Once fully proofed, make holes with your finger, dust semolina and sea salt.
6. "Plant" rosemary tips on the dough and bake directly on the baking stone, with steam. Open the exhaust once coloring starts.

製 法

1. 把迷迭香粗略切成細小塊。
2. 搓揉麵糰，在搓揉過程的最後 3 分鐘，加入鹽和碎迷迭香，續搓揉完成。
3. 初步發酵麵糰完成，放在已撒粗粒小麥粉的檯子上，量重量，讓麵糰自然平放成形。
4. 放在已撒粗粒小麥粉的發酵布上，發酵。
5. 一旦完成發酵完成，用手指在麵糰挖孔，撒上粗粒小麥粉和海鹽。
6. 在麵糰插上迷迭香尖，放入蒸氣，直接放在烘焙石上焗烤，當麵包轉色立即開啟排氣活門。

黑胡椒茴香脆條

Black Pepper & Fennel Taralli

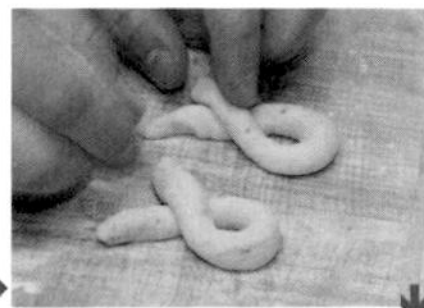
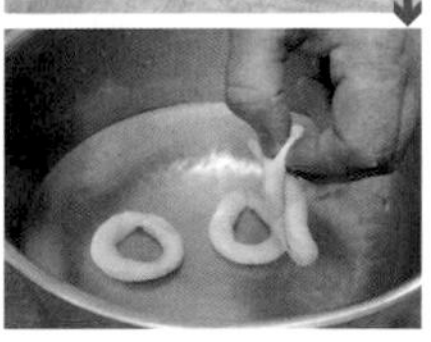

CHEF'S TIPS This classic "bread stick" from the south of Italy follows the same way of bagel's making, being boiled in water for a short while before baking. The taralli are usually shaped into closed rings, but I find other shapes attractive. If you find the texture too compact, you can add a little yeast or baking powder. The taralli can be flavored with any spices and can also be made in sweet version. These crispy sticks are great enjoyed with a glass of wine or with any soups.

麵包師建議 這傳統"麵包條"來自南義大利，跟貝哥麵包圈一樣，必須在烘焙前放在清水中烹煮片刻。脆餅一般造成近似環形，但其他造型也很吸引。如果覺得質感過於緊密，可加少許酵母或泡打粉。這些脆餅可與任何香料配搭，增加味道，也可做甜版本。可配一杯紅 / 白酒或任何湯品的小吃。

55°C	A: 8mins(S) B: 4mins(F)	23°C	30 mins	30 gm	no rest time	'U' shape sticks	none	200°C	200°C	35 mins

	INGREDIENTS	RATIO
500 gm	Flour type 00	100%
15 gm	Salt	3%
150 ml	White wine	30%
120 ml	Olive oil	24%
SQ	Black pepper, cracked	
SQ	Fennel seeds, dried	
	Dusting flour	
	Water for boiling	

	材料	比例
500 克	粉心麵粉	100%
15 克	鹽	3%
150 毫升	白酒	30%
120 毫升	橄欖油	24%
	黑胡椒適量（壓碎）	
	茴香籽適量（乾燥）	
	麵粉適量（撒面）	
	清水適量（燒滾）	

METHOD

1 Knead the dough with all ingredients and allow resting.
2 Cut the pieces and roll as bread stick of approximately 20 cm in length.
3 Boil water in a pot with a large tablespoon of salt.
4 Deep the pieces of dough in the hot water for a few minutes, until they float.
5 Using a strainer, take them out and place on a baking tray, forming different shapes.
6 Bake without steam and closed exhaust. Open the exhaust once coloring starts.
7 For very crispy and dry taralli, let them dry at a lower temperature for a longer time.

製 法

1 把所有材料混合，搓揉成麵糰，進行鬆弛。
2 麵糰切成小塊，滾成長條狀，每條長約 20 厘米。
3 將清水和 1 大湯匙鹽置於鍋中煮開。
4 將麵糰放到熱水數分鐘，直到麵糰浮起。
5 用篩網撈出，放在烤盤上，整出不同形狀。
6 不用放蒸氣於烤箱，關閉排氣活門烘烤，但當麵包開始轉色，立即開啟排氣活門。
7 想做到非常酥脆和較乾的南義大利半島脆餅，讓麵包條停留在已轉低溫的爐中長一點時間才取出。

63℃ | A: 8mins(S) B: 8mins(F) | 24℃ | 30 mins | 45 gm | no rest time | snail rolls | 45 mins | 210℃ | 190℃ | 25 mins

	INGREDIENTS	RATIO
250 gm	Flour type 65	100%
8 gm	Dry yeast	3.2%
5 gm	White sugar	2%
6 gm	Salt	2.4%
50 gm	Butter	20%
150 ml	Milk	60%
60 gm	Butter for folding	24%
SQ	Toasted pine nuts	
SQ	Fresh Basil	
	Sea salt	
	Dusting flour	

	材料	比例
250 克	高筋麵粉	100%
8 克	乾酵母	3.2%
5 克	細砂糖	2%
6 克	鹽	2.4%
50 克	奶油	20%
150 毫升	鮮奶	60%
60 克	奶油（夾心）	24%
	松子適量（已烘焙）	
	新鮮羅勒適量	
	海鹽適量	
	麵粉適量（撒面）	

METHOD

1. Knead the dough with all ingredients; add the salt 3 minutes before kneading ends.
2. Press the dough into a flat square of about 2 cm thickness and refrigerate for 3 hours.
3. Roll your folding butter between 2 pieces of baking paper to obtain a thin butter sheet.
4. Place the butter in the middle of the dough; fold the dough to cover the butter, like making croissant dough.
5. Roll at 2 cm and visually divide the dough in 3 parts and fold into 3 tiers.
6. Roll again the dough at 2 cm and repeat the folding 2 more times.
7. Wrap the dough and keep it in the fridge for 1 hour.
8. Roll the dough at 3 mm thickness and brush with a little water.
9. Sprinkle toasted pine nuts and fresh basil chiffonade; roll the dough into a long snail.
10. Relax the dough and cut into 40 gm portion. Place the dough into greased moulds and proof.
11. Sprinkle a few sea salt flakes and bake with steam and closed exhaust. Open the exhaust once coloring starts.

製 法

1 搓揉攪拌所有材料，並在搓揉過程的最後 3 分鐘，加入鹽續搓揉至完成。

5 把麵糰壓成扁的方形，厚約 2 厘米，置冰箱待 3 小時。

3 夾心奶油放在兩張烤盤紙的中央碾平，做出纖薄奶油片。

4 放在麵糰的中央，然後把麵糰摺疊奶油，做法如牛角麵包。

5 麵糰捲起 2 厘米，再按麵糰分成 3 份，覆摺成 3 層。

6 捲起麵糰 2 厘米，重複覆摺 2 次。

7 包起麵糰，置冰箱 1 小時。

8 把麵糰碾平約 3 毫米厚，刷點清水。

9 撒上已烘焙的松子和新鮮碎羅勒，捲成長螺紋狀。

10 麵糰鬆弛，切成 40 公克一份，放在已噴油的模型中發酵。

11 撒上海鹽片，放蒸氣於烤箱，關閉排氣活門才開始烘焙。當麵糰轉色，立即開啟排氣活門。

CHEF'S TIPS **To obtain an even coloration of your roll, remove the ring after ¾ of the baking time is past. When proofing this roll, ensure that the temperature of the place is not too high or else your butter will melt and the layering effect will be lost.**

麵包師建議 **為了讓麵糰的顏色平均，待烘焙時間已過了 ¾ 時，脫去外模。把此麵糰卷發酵，技巧是要確保置放的溫度不太高，否則奶油會融化，失去達到有層次感的效果。**

香脆松子羅勒卷
Crispy Pine Nuts Basil Rolls

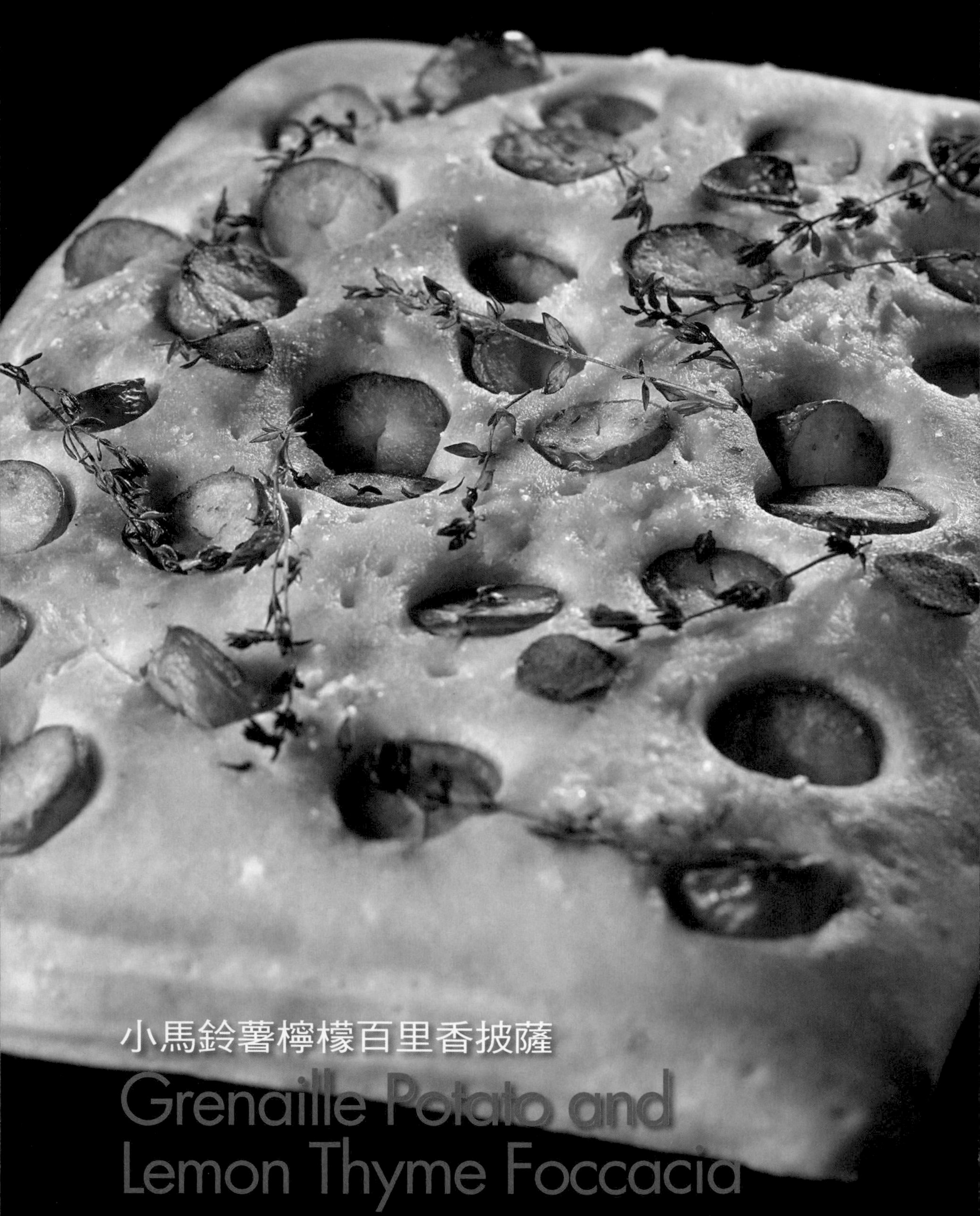

小馬鈴薯檸檬百里香披薩

Grenaille Potato and Lemon Thyme Foccacia

CHEF'S TIPS **The potatoes are baked without water with the purpose of not adding extra liquid to the dough. The addition of potatoes in the dough will bring an extra starch in the bread which will give a beautiful crust and perfect moisture to the bread. Lemon thyme, olive oil and sea salt are great flavors to go with smoked salmon and all those hearty Italian dishes.**

麵包師建議 **烘烤馬鈴薯無需用清水，目的是阻止過多水份加入麵糰內。於麵糰加入馬鈴薯可使麵包的澱粉質增加，造出一層香脆的外層，更可使麵包的水份恰到其份。此麵包的最佳吃法，便是加入檸檬百里香、橄欖油和海鹽於煙燻鮭魚或配合其他豐盛的義大利菜。**

64℃ | A: 8mins(S) B: 6mins(F) | 24℃ | 1 hr | tray size | none | in tray | 45 mins | 225℃ | 200℃ | 40 mins

	INGREDIENTS	RATIO
375 gm	Flour type 45	100%
150 ml	Fresh milk	40%
150 ml	Water	40%
40 ml	Olive Oil	10.67%
12 gm	Salt	3.2%
15 gm	Dry yeast	4%
375 gm	Baked potatoes	100%
SQ	Baked grenaille potatoes	
SQ	Fresh lemon thyme	
SQ	Sea salt	
SQ	Olive oil	
	Dusting flour	

	材料	比例
375 克	低筋麵粉	100%
150 毫升	鮮奶	40%
150 毫升	清水	40%
40 毫升	橄欖油	10.67%
12 克	鹽	3.2%
15 克	乾酵母	4%
375 克	烤馬鈴薯	100%
	烤小馬鈴薯適量	
	新鮮檸檬百里香適量	
	海鹽適量	
	橄欖油適量	
	麵粉適量（撒面）	

METHOD

1. Bake the unpeeled baking potatoes in a tray covered by aluminum foil at 200?C for about 1 hour, until fully cooked. Once cold, peel them and crush them in chunky puree.
2. Cook the grenaille potatoes with their skin and let cool. Cut into 1cm thick slices.
3. Knead the dough and add the salt and the crushed potatoes 3 minutes before kneading ends.
4. After the bulk fermentation, roll the dough in an oiled tray and proof.
5. Once fully proofed, puncture holes with your fingers randomly and brush olive oil.
6. Place the grenaille potatoes slices on the dough, pressing them a little, and sprinkle sea salt and lemon thyme.
7. Bake with steam and closed exhaust. Open the exhaust once coloring starts.

製 法

1. 把馬鈴薯連皮放在烤盤，蓋上鋁箔紙，以爐溫 200℃烤 1 小時，直至完全熟透。趁熱剝去外皮，壓成粗薯泥狀。
2. 放涼烤熟的連皮小馬鈴薯，切成 1 厘米厚片。
3. 搓揉麵糰，在搓揉過程的最後 3 分鐘，加入鹽和薯泥，續至搓揉完成。
4. 初步發酵麵糰完成，擀平麵糰，放進已刷油的烤盤，進行發酵。
5. 麵糰完全發酵後，在麵糰隨意以手指刺上幾個小孔，然後刷上橄欖油。
6. 放小薯片於麵糰上，輕輕按壓，撒上海鹽和檸檬百里香。
7. 放蒸氣於烤爐內，關閉排氣活門，然後烘焙，待麵糰轉色，立即開啟排氣活門。

培根香草麥穗
Bacon & Herbs Épi

CHEF'S TIPS **Roll the baguette in a little flour once finished to shape to give it a rustic look. This baguette is convivial and attractive to eat with friends, pulling apart each single roll. To obtain the characteristic shape of the ?pi, ensure to not over proof your baguette before cutting.**

麵包師建議 **在長形棍狀麵糰上撒少許麵粉，繼而碾擀麵糰，可營造出粗糙感。把它分割成小卷，更是宴會上跟朋友歡聚，不可或缺的小吃，非常吸引人。**

64℃ | A: 10mins(S) B: 8mins(F) | 26℃ | 1.5 hrs | 350 gm | 15 mins | pi baguette | 1.5 hrs | 235℃ | 215℃ | 35 mins

	INGREDIENTS	RATIO
	Poolish	
200 gm	Flour type 65	
200 ml	Water	
3 gm	Dry yeast	
	Dough	
400 gm	Flour type 65	100%
400 gm	Ripe poolish	100%
2 gm	Dry yeast	0.5%
15 gm	Salt	3.75%
250 ml	Water	62.5%
35 gm	Sauteed bacon	8.75%
4 gm	Mixed dry herbs	1%
	Dusting flour	

METHOD

1. Knead the dough and add the salt 3 minutes before kneading ends.
2. Add the bacon and herbs in slow speed, once the kneading process is completed.
3. Allow resting and weigh the dough. Pre-shape in baguette.
4. Once rested, shape the dough into baguette and proof on a floured cloth.
5. Cut in épi using a metal dough scraper and bake on tray or directly on stone.
6. Bake with steam and closed exhaust. Open the exhaust once coloring starts.

	材料	比例
	速成麵種	
200 克	高筋麵粉	
200 毫升	清水	
3 克	乾酵母	
	麵糰	
400 克	高筋麵粉	100%
400 克	熟速成麵種	100%
2 克	乾酵母	0.5%
15 克	鹽	3.75%
250 毫升	清水	62.5%
35 克	炒香培根	8.75%
4 克	乾香草	1%
	麵粉適量（撒面）	

製 法

1. 搓揉麵糰，在搓揉過程的最後 3 分鐘，加入鹽續搓揉完成。
2. 搓揉過程一旦完成，以慢速加入培根和香草。
3. 鬆弛麵糰，量重量，把麵糰初整型，整成長棍狀。
4. 當鬆弛完成，再次捲成長棍狀，放在已撒麵粉的發酵布上發酵。
5. 麵包剪成麥穗狀，放在烤盤烘烤或直接放在烘焙石均可。
6. 放蒸氣於烤箱，關閉排氣活門，然後烘焙，待麵包轉色，立即開啟排氣活門。

帕馬火腿芝麻菜披薩

Arugula & Parma Ham Pizza

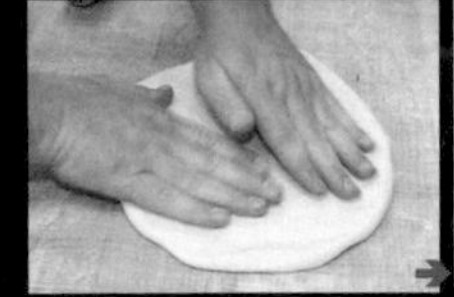  

CHEF'S TIPS The thickness of the dough depends on your liking. You may do it thin and crispy or thicker and more bread-like. To add a little luxurious touch to your pizza, add a drizzle of truffle oil or even freshly slices truffles on top. The flour type 00 suits best the purpose, but other flour are also fine to use. A thin pizza should be baked in a very hot oven for a short time, giving these characteristic dark edges to the pizza.

麵包師建議 麵糰的厚薄，適隨尊便。你可以做成薄脆或厚如麵包。想添加豪華感，可在薄餅上滴上少許黑松露油甚至放上黑松露片。粉心的麵粉最適合不過，但其他優質麵粉也可以。想令薄餅有脆薄的效果，必須以短時間在非常高溫的烤箱烘烤，這樣可以使薄餅周邊呈深色特點。

     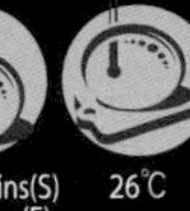  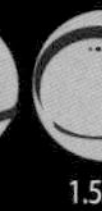           

64℃ | A: 10mins(S) B: 8mins(F) | 26℃ | 1.5 hrs | 200 gm | 1 hr | flat round | none | 270℃ | 270℃ | approximately 12 mins

	INGREDIENTS	RATIO
250 gm	Flour type 00	100%
150 ml	Water	60%
4 gm	Dry yeast	1.6%
12 gm	Sea salt	4.8%
SQ	Arugula	
SQ	Parma ham slice	
SQ	Parmesan cheese	
SQ	Tomato sauce	
SQ	Fresh oregano	
SQ	Olive oil	
SQ	Black pepper	
SQ	Mozzarella cheese	
	Dusting flour	

	材料	比例
250 克	粉心麵粉	100%
150 毫升	清水	60%
4 克	乾酵母	1.6%
12 克	海鹽	4.8%
	芝麻菜（火箭菜）適量	
	帕爾馬火腿片適量	
	帕爾馬起司適量	
	番茄醬適量	
	新鮮奧勒岡葉適量	
	橄欖油適量	
	黑胡椒適量	
	馬茲瑞拉起司適量	
	麵粉適量（撒面）	

METHOD

1. Knead the dough and add the salt 3 minutes before kneading ends.
2. After the bulk fermentation, weigh and form balls of dough.
3. Cover with a plastic film and allow resting at room temperature.
4. During the resting time, prepare all the ingredients needed for the topping.
5. Roll your dough on fine semolina at about 0.5 to 1 cm in thickness.
6. Ensure to always have enough semolina to avoid the dough from sticking to the surface.
7. Spread the tomato sauce, leaving the edges free of sauce. Sprinkle the shredded mozzarella cheese.
8. Bake in a very hot oven directly on your baking stone without steam and open exhaust.
9. Once baked and while hot, add all the other ingredients randomly, finishing the top with the arugula.

製 法

1. 搓拌麵糰，在搓揉過程的最後 3 分鐘，加入鹽續搓揉完成。
2. 初步發酵麵糰完成後，量重量，整型成球狀麵糰。
3. 蓋上保鮮膜，置室溫下進行鬆弛。
4. 在鬆弛時間，準備鋪面的所需材料。
5. 麵糰放在粗粒麥粉，碾擀成 0.5 至 1 厘米厚的圓形。
6. 確保時常有足夠粗粒小麥粉，避免黏貼在工作枱上。
7. 抹上番茄醬，讓四角不沾醬汁，撒上馬茲瑞拉起司碎。
8. 不用放入蒸氣，開啟排氣活門，直接放在高溫的烤箱烘焙石烘烤。
9. 當烘焙完成，趁熱把其他剩餘材料，隨意加入在薄餅上，完成後放上芝麻菜。

CHEF'S TIPS Eating this bread cold will not be as pleasant as warm as the butter topping will be set. Use a lot of garlic and parsley to make the flavors stronger. In Russian classic cooking, the pampushka is usually served with Borsch soup. If you want to have your bread with more butter, don't apply the egg wash before baking. That way, it'll be less shiny, but butter will go deeper in the bread crumb.

麵包師建議 這麵包凍涼才享用，不及仍處暖和時來得好，因為奶油會在麵包裡面凝固。選用大量蒜茸和洋香菜可增強味道。這款軟麵包在俄羅斯的傳統烹飪是與羅宋湯伴吃。如果你想麵包多點奶油，烘焙前就不要刷蛋液。如此做法，麵包的光澤會減少，但奶油會滲於麵包內。

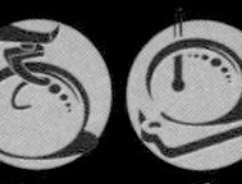

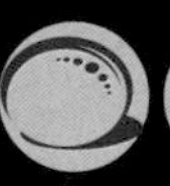

64℃ | A: 10mins(S) B: 6mins(F) | 25℃ | 1 hr | 30 gm | 15 mins | bunch of round | 45 mins | 215℃ | 200℃ | 35 mins

INGREDIENTS

	INGREDIENTS	RATIO
500 gm	Flour type 45	100%
140 gm	Butter	28%
50 gm	White sugar	10%
5 gm	Salt	1%
10 gm	Dry yeast	2%
275 ml	Fresh milk	55%
60 gm	Butter, melted	
1 head	Garlic, fresh, chopped	
SQ	Parsley, fresh, chopped	
	Dusting flour	
	Egg wash	

	材料	比例
500 克	低筋麵粉	100%
140 克	奶油	28%
50 克	細砂糖	10%
5 克	鹽	1%
10 克	乾酵母	2%
275 毫升	鮮奶	55%
60 克	奶油（融化）	
	新鮮蒜頭 1 球（剁碎）	
	新鮮洋香菜適量（剁碎）	
	麵粉適量（撒面）	
	蛋液適量（掃面）	

METHOD

1. Melt the butter in a pot and just before boiling point, add the chopped raw garlic and the parsley. Let it sit in a warm place to infuse the flavors.
2. Knead the dough and add the salt 3 minutes before kneading ends.
3. After the bulk fermentation, weigh and allow resting.
4. Form the round rolls and place them, touching each others, in an oiled baking tray.
5. Proof the dough and brush egg wash once.
6. Bake with steam and closed exhaust. Open the exhaust once coloring starts.
7. Once baked and while hot, brush the garlic butter a few times on the bread and enjoy while still warm.

製 法

1. 奶油放小鍋內融化，在沸騰前加入鮮蒜頭碎和洋香菜碎，放在溫暖的地方讓其味道泡出。
2. 搓揉麵糰，在搓揉過程的最後 3 分鐘，加鹽續至搓揉完成。
3. 初步發酵麵糰後，量重量和進行鬆弛。
4. 捲成圓卷，把圓卷的邊沿互相緊貼，置放在已塗油的烤盤。
5. 發酵麵糰，待完成後刷上蛋液。
6. 放入蒸氣於爐內，關閉排氣活門，然後烘烤，待麵包開始轉色，立即開啟排氣活門。
7. 麵包出爐後，趁熱抹上香蒜奶油等帶片刻，趁熱享用。

蒜蓉香草俄羅斯麵包

Russian Pampushka

紅甜椒瑞士起司卷
Red Bell Pepper and Bagnes Cheese Rolls

62℃ | A: 10mins(S) B: 5mins(F) | 25℃ | 1 hr | 40 gm | 15 mins | rolls | 45 mins | 235℃ | 210℃ | 30 mins

	INGREDIENTS	RATIO
380 gm	Flour type 65	100%
75 gm	Rye flour	19.74%
6 gm	Dry yeast	1.32%
12 gm	Salt	2.64%
280 ml	Water	61.54%
115 gm	Bagnes cheese	25.27%
65 gm	Red bell pepper	14.29%
	Dusting flour	

	材料	比例
380 克	高筋麵粉	100%
75 克	裸麥麵粉	19.74%
6 克	乾酵母	1.32%
12 克	鹽	2.64%
280 毫升	清水	61.54%
115 克	瑞士巴山谷起司	25.27%
65 克	紅甜椒	14.29%
	麵粉適量（撒面）	

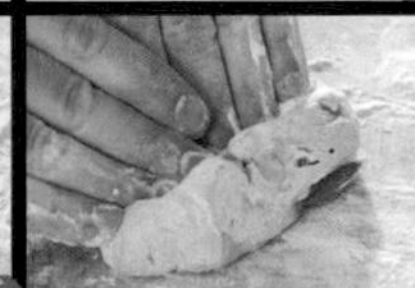

CHEF'S TIPS The Bagnes cheese from the Valais region in Switzerland. It is recognized as the finest cheese for Raclette. This type of semi-hard cheese is perfect to bake in bread and pairs nicely with roasted bell pepper. If not available, you can substitute Bagnes cheese with other semi-hard cheeses. Cheese will take a fast brown coloring while baking; therefore, at about 10 minutes from the end of baking, you might to keep your oven's door ajar. Freshly baked and served warm, this bread and its melting pieces of cheese will be the perfect fit for an arugula salad with ripe cherry tomatoes.

麵包師建議 瑞士巴谷山起思源自瑞士瓦萊州。它被公認為最佳的 Racletle 起司。這半硬起司放在麵包內烘焙或配已烤甜椒是非常完美組合。要是沒有這半硬起司，你可用瑞士巴谷起司代替。起司在烘焙時很快變棕褐色，所以在完成烘焙前 10 分鐘，你可以半開爐門。剛出爐的麵包上有正融化的起司，需趁熱享用。它與芝麻菜沙拉和成熟了的櫻桃番茄十分匹配。

METHOD

1. Prepare the bell peppers; on a flame stove, burn until the skin in black color. Once fully burnt, wrap the pepper in a plastic film and let it rest for 15 minutes. You can then peel the pepper easily without using water bath. Water would reduce the flavors of the bell pepper. Then, empty the middle and cut the pepper in rough pieces of about 2 to 3 cm.
2. Dice the Bagnes cheese into cubes of about 1 cm.
3. Knead the dough and add salt 3 minutes before the end of mixing. Once the kneading is done, add the diced cheese and bell pepper in slow speed.
4. After the bulk time, weigh the dough and pre-shape the rolls.
5. Brush a little oil on the bottom of the roll and give the final roll shape to the dough.
6. Proof with the oiled surface facing downward, on a floured cloth.
7. Once fully proofed, turn the rolls over and bake with a little steam. Open the steam exhaust once coloration starts.

製 法

1. 把紅甜椒放在瓦斯爐上燒至外皮焦黑。一旦完成燒焦，用塑膠袋包裹紅椒，讓其停放 15 分鐘後非常容易地去掉外皮，無需浸泡於水中，因清水會降低紅椒的味道，然後去籽，粗略地切小塊約 2 至 3 厘米。
2. 瑞士巴山谷起司切成 1 厘米丁粒。
3. 搓揉麵糰，在搓揉過程的最後 3 分鐘，加入鹽續搓揉至完成。麵糰搓揉完成，以慢速加入起司粒和紅甜椒塊。
4. 初步發酵麵糰完成後，量重量，初整型。
5. 在小麵糰底部刷點油，最後整型。
6. 發酵後，把塗了油的麵糰翻轉向下，然後放在已撒麵粉的發酵布上最後發酵。
7. 一旦發酵完成，翻轉小麵糰，放入少量蒸氣，然後烘焙，待麵包轉色，立即開啟排氣活門。

61℃ | A: 8mins(S) B: 3mins(F) | 23℃ | 30 mins | - | none | flat crisp | 1 hr | 170℃ | 170℃ | 35 mins

	INGREDIENTS	RATIO
520 gm	Flour type 65	100%
260 gm	Flour type 45	50%
10 gm	White sugar	1.28%
20 gm	Salt	2.56%
12 gm	Dry yeast	1.54%
150 ml	Fresh milk	19.23%
190 ml	Water	24.36%
110 gm	Butter for folding	14.1%
SQ	Espelette Pepper	
	Dusting flour	

	材料	比例
520 克	高筋麵粉	100%
260 克	低筋麵粉	50%
10 克	細砂糖	1.28%
20 克	鹽	2.56%
12 克	乾酵母	1.54%
150 毫升	鮮奶	19.23%
190 毫升	清水	24.36%
110 克	奶油 （夾心）	14.1%
	法式紅椒粉適量	
	麵粉適量（撒面）	

METHOD

1 Knead the dough with a not so strong gluten network.
2 After the bulk fermentation, flatten your dough to 2 cm thick and store in the chiller for about 2 hours, covered with a plastic film.
3 Roll your folding butter between 2 pieces of baking paper to obtain a thin butter sheet.
4 Place the butter in the middle of the dough and fold the dough to cover the butter, like in making croissant dough.
5 Roll at 2 cm and visually divide the dough in 3 parts and fold into 3 tiers.
6 Roll again the dough at 2 cm and repeat the folding 2 more times.
7 Wrap the dough and keep it in the fridge for 1 hour.
8 Roll the dough paper-thin and lay it on a baking tray with baking paper.
9 Allow resting for half an hour and lightly brush with water. Sprinkle Espelette pepper.
10 You may cut it now, or leave it as a whole sheet to be broken into rough pieces once baked.
11 Bake without steam and closed exhaust. Open the exhaust once coloring starts.

製 法

1 攪拌搓揉麵糰，麵筋不要太堅韌。
2 發酵後，擀平麵糰厚約 2 厘米，蓋上保鮮膜，放入冰箱冷藏 2 小時。
3 夾心奶油放在 2 張烘焙紙的中央，擀平成薄片。
4 放在麵糰的中央，覆蓋麵糰，如牛角麵包做法。
5 碾擀成 2 厘米，把麵糰分 3 份，摺疊成 3 層。
6 碾擀成 2 厘米麵糰，重複再折 2 次。
7 包裹麵糰，放入冰箱貯藏 1 小時。
8 碾壓麵糰如薄紙狀，排放在已鋪烘焙紙的烤盤上。
9 鬆弛約半小時，薄薄刷上清水，撒上紅椒粉。
10 此時可以切割麵糰，或整片等待烘焙後才粗略地弄成碎片狀。
11 無需放入蒸氣，關閉排氣活門，然後烘焙，待脆片開始轉色，立即開啟排氣活室。

CHEF'S TIPS The thinner the better! You can also add some sea salt or other spices, up to your taste buds. But beware, once you start eating these Lavosh, you won't be able to stop! The Lavosh won't rise during the fermentation time, but the bread will eventually generate a few bubbles during the baking process. The long resting time is simply to prevent the dough from retracting itself too much, especially when cutting shapes such as squares or rectangles, to keep their shape as sharp as possible.

麵包師建議 脆餅做法是越纖薄越好。按你的口味，可加點海鹽或香辛料，但要小心，因為你開始享用這美味的薄脆便停不了。薄脆會在發酵時膨脹，但麵包會在烘焙過程產生少量氣泡。長時間進行鬆弛過程，能防止麵糰縮回原狀，特別是在切割方形或三角形時，可令其周邊盡量保持尖削的效果。

法國紅辣椒薄脆

Piment d'Espelette Lavosh

普羅旺斯面具
Provencal Fougasse

CHEF'S TIPS Fougasse is thin flat bread from southern France that resembles Foccacia bread. You can add other ingredients to the dough or use different toppings before baking. Another version of the Fougasse is with stuffing inside the dough.

麵包師建議 Fougasse 面具麵包是南部法國的 Foccacia，屬香草麵包類。你可以在麵糰加入其他材料烘焙或選用不同配料也可。另一個 Fougasse 的版本是在麵包內釀入不同餡料，同樣美味。

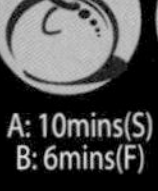

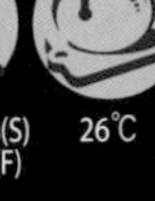
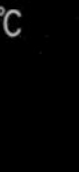

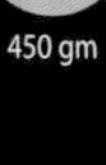

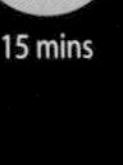

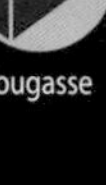

64℃ | A: 10mins(S) B: 6mins(F) | 26℃ | 1.5 hrs | 450 gm | 15 mins | fougasse | 45 mins | 235℃ | 225℃ | 40 mins

INGREDIENTS

	INGREDIENTS	RATIO
	Starter	
150 gm	Flour type 65	
180 ml	Water	
5 gm	Dry yeast	
5 gm	Sugar	
	Dough	
550 gm	Flour type 65	100%
350 ml	Water	63.64%
330 gm	Starter	60%
7 gm	Dry yeast	1.27%
20 gm	Salt	3.64%
20 ml	Fresh milk	3.64%
20 ml	Olive oil	3.64%
6 gm	Provence herbs, dried	1.09%
50 gm	Chopped sun dry tomato	9.09%
30 gm	Black olives	5.45%
30 gm	Bayonne Ham	5.45%

	材料	比例
	麵種	
150 克	高筋麵粉	
180 毫升	清水	
5 克	乾酵母	
5 克	細砂糖	
	麵糰	
550 克	高筋麵粉	100%
350 毫升	清水	63.64%
330 克	麵種	60%
7 克	乾酵母	1.27%
20 克	鹽	3.64%
20 毫升	鮮奶	3.64%
20 毫升	橄欖油	3.64%
6 克	普羅旺斯乾香草	1.09%
50 克	風乾番茄	9.09%
30 克	黑橄欖	5.45%
30 克	Bayonne 火腿	5.45%

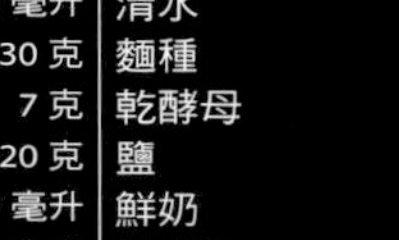
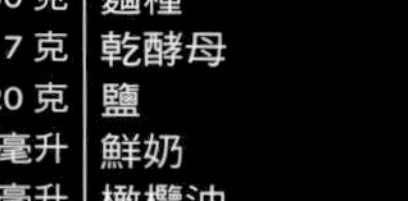
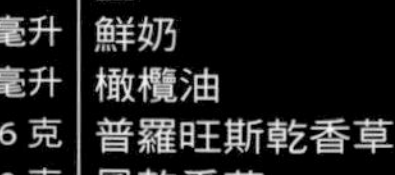
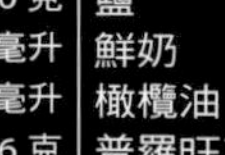
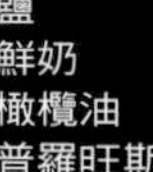
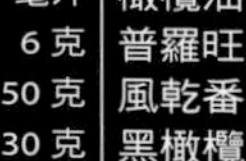

METHOD

1. Mix the starter and let it ferment for 4 to 5 hours at room temperature.
2. Knead the dough and add the salt 3 minutes before kneading ends.
3. Add the tomatoes, olives, herbs and Bayonne ham pieces in slow speed.
4. After the resting times, roll the dough in large rectangle slabs.
5. Give one long cut in the middle and 3 cuts on each side of it. Open the cuts and place on a baking tray.
6. Bake with steam until golden brown. Open the exhaust once coloring starts.

製 法

1. 麵種材料混合，待發酵放置室溫約 4~5 小時。
2. 搓揉麵糰，在搓揉過程的最後 3 分鐘，加入鹽續搓揉完成。
3. 以慢速加入番茄、橄欖、香草和 Bayonne 火腿片。
4. 鬆弛完成後，碾擀麵糰呈長方形厚片。
5. 在中間處沿麵糰長度切割一刀，每邊分別再切 3 刀，揭開切口，放入烤盤。
6. 放入蒸氣烘烤至金黃色，待麵包開始轉色，立即開啟排氣活門。

CHEF'S TIPS **Semi freezing the dough allows you to obtain very sharp shape. Ensure to use freshly chopped garlic and parsley, as well, don't mix these 2 ingredients for too long. Both, the garlic alliinase and the parsley chlorophyll have been scientifically identified as substances affecting the gluten strength and thus, the crumb of your breads.**

麵包師建議 **麵糰半冷藏可做出非常尖銳的形狀。確保採用新鮮蒜泥和洋香菜碎，並且不要把它們混合太久。經科學驗証，蒜頭的蒜氨酸跟洋香菜的葉綠素經長時間混合後，它會影響到麵筋的Q度，以至於麵包內層的質感。**

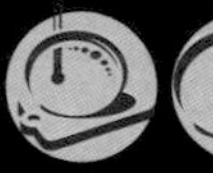

64℃ | A: 10mins(S) B: 6mins(F) | 26℃ | 1 hr | 25 gm | 30 mins | sharp needle | 45 mins | 200℃ | 185℃ | 20 mins

	INGREDIENTS	RATIO
	Starter	
150 gm	Flour type 55	100%
40 gm	Butter	26.67%
40 gm	Sugar	26.67%
40 gm	Egg Yolk	26.67%
45 gm	Whole Egg	30%
12 gm	Dry yeast	8%
100 ml	Water	66.67%
	Dough	
150 gm	Flour type 55	100%
15 gm	Milk powder	10%
7 gm	Salt	4.67%
20 gm	Freshly chopped garlic	13.33%
5 gm	Freshly chopped parsley	3.33%
	Egg wash	

	材料	比例
	麵種	
150 克	中筋麵粉	100%
40 克	奶油	26.67%
40 克	細砂糖	26.67%
40 克	蛋黃	26.67%
45 克	全蛋	30%
12 克	乾酵母	8%
100 毫升	清水	66.67%
	麵糰	
150 克	中筋麵粉	100%
15 克	奶粉	10%
7 克	鹽	4.67%
20 克	新鮮大蒜（切碎）	13.33%
5 克	新鮮洋香菜	3.33%
	蛋液適量（刷面）	

METHOD

1. Knead the starter and let it ferment for 1 hour at room temperature.
2. Knead the dough and add the salt 3 minutes before kneading ends.
3. Add the garlic and parsley at the end of kneading in slow speed.
4. After the bulk time, roll the dough in 1 cm thick rectangle of about 15cm in width.
5. Lay the dough on a floured tray, cover with a plastic film and store in the freezer for 30 minutes.
6. Place the dough on a floured cutting board and cut the long triangle needle.
7. Put them on a baking tray and egg wash once; proof.
8. Once fully proofed, egg wash once more and bake with steam. Open the exhaust once coloring starts.

製 法

1. 把所有材料混合搓揉成麵種，讓其置於室溫下發酵 1 小時。
2. 搓揉麵糰，在搓揉過程的最後 3 分鐘，加入鹽續搓揉至完成。
3. 以慢速拌入大蒜和洋香菜。
4. 發酵麵糰後，碾擀麵糰成長方形，約 1 厘米厚 x 15 厘米寬。
5. 放已撒麵粉的烤盤，蓋上保鮮膜，放冰箱冷藏 30 分鐘。
6. 把麵糰放在已撒麵粉的板上，切成長三角針形。
7. 放入烤盤，刷上蛋液，進行最後發酵。
8. 當發酵完成，再刷一次蛋液，放入蒸氣，然後烘焙，待麵包開始轉色，立即開啟排氣活門。

大蒜洋香菜麵包
Garlic Parsley Brioche Needle

培根葡萄乾小法國

Cider Raisins, Smoked Bacon and Coriander Rolls

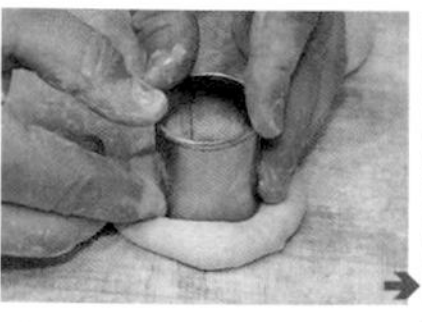
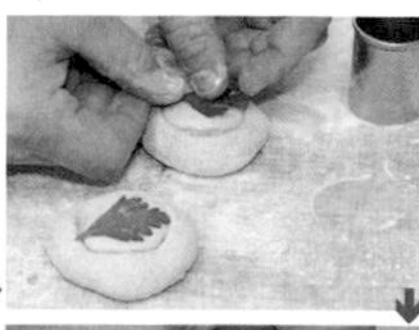
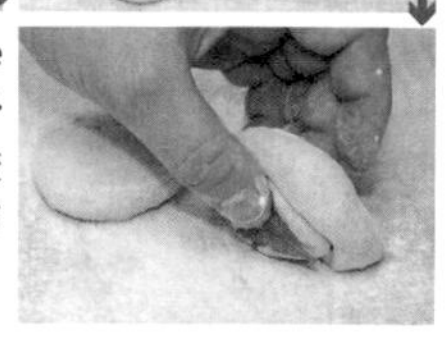

CHEF'S TIPS **Use light rye flour to dust the table where you proof your rolls, it'll give you a better contrast and flavor. For an added fruity flavor and texture, add one apple (type Jonagold or Gala) in rough dice throughout the dough; very nice with bacon, raisins, coriander and cider.**

麵包師建議 **選用粿麥麵粉撒在已發酵的小麵卷的檯上，這樣會給予較好的對比和味道。添加果味和改善質地，可加一個蘋果（金蘋果或節日蘋果）的粗丁粒，混合麵糰，它與培根、葡萄乾、香菜和蘋果酒，十分匹配。**

64℃ | A: 10mins(S) B: 8mins(F) | 25℃ | 2 hrs | 50 gm | 15 mins | round rolls | 2 hrs | 230℃ | 210℃ | 25 mins

	INGREDIENTS	RATIO
	Poolish	
80 gm	Flour type 65	
100 ml	Apple Cider	
2 gm	Dry yeast	
	Dough	
320 gm	Flour type 65	100%
130 gm	Rye flour type 770	-
200 gm	Above ripe poolish	44%
1 gm	Dry yeast	~
12 gm	Sea Salt	2.7%
120 gm	Raisins soaked in cider	26.7%
12 gm	Coriander seeds crushed	2.7%
120 gm	Smoked bacon sautéed	26.7%
350 ml	Cider	77.8%
	Fresh coriander leaves	

	材料	比例
	速成麵種	
80 克	高筋麵粉	
100 毫升	蘋果酒	
2 克	乾酵母	
	麵糰	
320 克	高筋麵粉	100%
130 克	粿麥麵粉	-
200 克	速成麵種	44%
1 克	乾酵母	~
12 克	海鹽	2.7%
120 克	酒漬葡萄乾	26.7%
12 克	香菜籽碎	2.7%
120 克	炒熟培根	26.7%
350 毫升	蘋果酒	77.8%
	新鮮香菜葉適量	

METHOD

1 Prepare the poolish by mixing the ingredients with a whisk and keep it at room temperature to ferment for 4 hours.

2 Soak your raisins in enough cider to be submerged for the same amount of time.

3 Cut your smoked bacon in long strip and saut?ed it dark brown.

4 Mix and knead your dough with all the ingredients except salt, raisins, bacon and coriander leaves. Add the salt 3 minutes before kneading ends.

5 Once the dough is kneaded, add the raisins and bacon in slow speed.

6 Once rested, weigh the rolls, pre-shape and allow resting on the bench.

7 Shape round rolls, cut a ring through w of the roll. Brush a little water.

8 Stick a leaf of fresh coriander, and place them up-side-down on a floured surface for proofing.

9 Once proofed, flip them over and bake them with steam until golden brown.

製 法

1 用打蛋器拌勻速成麵種材料，放室溫下發酵 4 小時。

2 把葡萄乾完全浸泡於足夠的蘋果酒內 4 小時。

3 把培根切成長條，炒至深褐色。

4 混合所有材料，除了鹽、酒浸葡萄、培根和香菜葉外，搓揉麵糰，在搓揉過程的最後 3 分鐘，加入鹽搓揉至完成。

5 麵糰搓揉一旦完成，以慢速拌入酒浸葡萄和培根。

6 鬆弛完成後，秤出小麵卷的重量，初整型，放在木板上鬆弛。

7 滾成圓形，壓出 w 小麵糰，刷點清水。

8 貼上新鮮香菜葉，反轉麵糰，置在已撒上麵粉的工作檯上最後發酵。

9 當發酵過程完成，反轉麵糰，放入蒸氣於烤箱內，烘焙至金黃色。

Almost

跨世紀麵包

Untouchable

法式長麵包
French Baguette

CHEF'S TIPS **Using a poolish is a great way to produce flavorful and crusty bread over a medium time of fermentation.**
When making the poolish, use room temperature flour and water. Using an excessively warm water will eventually result in an over proofed poolish. It will also not produce the expected fermentation in the final dough. The poolish can be made over longer period of time kept to ferment at 5˚C. Longer time of fermentation means better flavors and conservation.

麵包師建議 **一般中的發酵時間而言，使用速成麵種是最好的方法去製作出非常有風味而外層又香脆和麵包。速成麵種，只要用上置室溫下的麵粉和清水便可。用上額外的溫水，最終結果是過度發酵速成麵種，更不能做出預期效果。速成麵種可以長時間貯存在溫度 5˚C 而發酵，用較長時間的用意是令其味道更好和可延長保存期。**

64˚C | A: 10mins(S) B: 8mins(F) | 26˚C | 1.5 hr | 350 gm | 15 mins | baguette | 1 hr | 235˚C | 215˚C | 35 mins

	INGREDIENTS	RATIO
	Starter	
90 gm	Flour type 65	
90 ml	Water	
3 gm	Dry yeast	
	Dough	
180 gm	Flour type 65	100%
110 ml	Water	61.11%
2 gm	Dry yeast	1.11%
180 gm	Starter (all of above)	100%
6 gm	Salt	3.33%
	Dusting flour	

	材料	比例
	原始麵種	
90 克	高筋麵粉	
90 毫升	清水	
3 克	乾酵母	
	麵糰	
180 克	高筋麵粉	100%
110 毫升	清水	61.11%
2 克	乾酵母	1.11%
180 克	麵種	100%
6 克	鹽	3.33%
	麵粉適量（撒面）	

METHOD

1. Mix the starter in smooth dough and let it ferment at room temperature for 5 hours.
2. Knead the dough and add the salt 3 minutes before kneading ends.
3. Allow resting and weigh the dough. Pre-shape baguette and let rest on bench.
4. Once rested, shape the baguette and end the shaping with a slight touch of flour.
5. Once proofed, score the baguette 4 to 5 times and bake directly on the baking stone.
6. Bake with steam and closed exhaust. Open the exhaust once coloring starts.

製 法

1. 混合麵種材料成光滑麵糰，置室溫下發酵 5 小時。
2. 搓揉麵糰，在搓揉過程的最後 3 分鐘，加鹽續搓揉至完成。
3. 鬆弛麵糰，量重量，整型成法國麵包的長棒形，放在木板上鬆弛。
4. 當鬆弛完成，修飾造型，在整型後撒上少許麵粉發酵。
5. 發酵一旦完成，在法國麵包上用刀刻紋 4~5 次，置在烘焙石直接烘烤。
6. 放入蒸氣，關閉排氣活門，然後烘焙，待麵包開始轉色，立即開啟排氣活門。

蕎麥枕頭麵包
Buckwheat Pillow Bread

CHEF'S TIPS Ideally, fermented dough is a piece of dough you have left from the day before and stored in the fridge. If you don' t have it on hand, you can produce a small dough. To produce the fermented dough, you can use 150 gm of flour type 65 and 100 ml of water with 2 or 3 grams of yeast, knead it, let it ferment for 2 hours at room temperature and leave it in the fridge until the next day before using.

Cut a stencil in a soft cardboard or paper that can adapt to the delicate surface of the proofed bread. Ensure the surface to be dry enough not to have the stencil sticking to the dough.

麵包師建議 最理想的已發酵麵糰是指麵糰在預早一天準備好，並貯放在冰箱中備用。如果沒有預先準備，你可以製造少量麵糰。取用高筋麵粉約 150 克、清水 100 毫升、2 或 3 克酵母，搓揉成糰，待室溫下發酵 2 小時，放進冰箱就可在翌日使用。

用軟身的厚紙板或紙張剪出模版，紙質必須適合細緻表面的已發酵麵糰，並確保模版表面乾燥，以防止貼在一起。

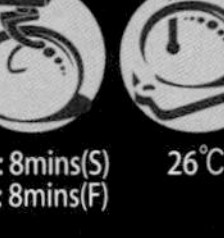

64℃ | A: 8mins(S) B: 8mins(F) | 26℃ | 18 hrs at 5℃ | 800 gm | 3 hrs | rectangle load | 1 hr | 225℃ | 200℃ | 50 mins

	INGREDIENTS	RATIO
250 gm	Flour type 65	100%
75 gm	Whole wheat flour	
50 gm	Buckwheat flour	
150 gm	Fermented dough	40%
3 gm	Dry yeast	0.8%
20 gm	Wheat germs	5.33%
15 gm	Salt	4%
350 ml	Water	93.33%
SQ	Dusting flour	

	材料	比例
250 克	高筋麵粉	100%
75 克	全麥麵粉	
50 克	蕎麥麵粉	
150 克	發酵麵糰	40%
3 克	乾酵母	0.8%
20 克	小麥胚芽	5.33%
15 克	鹽	4%
350 毫升	清水	93.33%
	麵粉適量（撒面）	

METHOD

1. Knead the dough and add the salt 3 minutes before kneading ends.
2. After the long bulk fermentation, let the dough sit at room temperature for 3 hours.
3. Flatten the dough and fold it in 3 layers once, like for croissant dough.
4. Allow resting for 10 minutes and flatten it again to about 4 cm in thickness.
5. Cut the rectangles of dough and place on a floured cloth to proof.
6. Prepare your stencil and once proof, place the stencil on the bread and dust with rye four.
7. Bake with steam and closed exhaust. Open the exhaust once coloring starts.

製 法

1. 搓揉麵糰，在搓揉過程的最後 3 分鐘，加鹽續搓揉至完成。
2. 經過長時間發酵後，讓麵糰置室溫下待 3 小時。
3. 碾平麵糰，複摺成 3 層，如牛角包做法。
4. 鬆弛麵糰 10 分鐘，再次碾平，約 4 厘米厚。
5. 切割成長方形麵糰，放在已撒麵粉的發酵布上發酵。
6. 準備模版，待麵糰發酵後放上，再撒上粿麥麵粉。
7. 放入蒸氣入烤爐，關閉排氣活門，然後烘焙，待麵包開始轉色，立即開啟排氣活門。

瑞士辮子麵包

Petite Tresse au Beurre

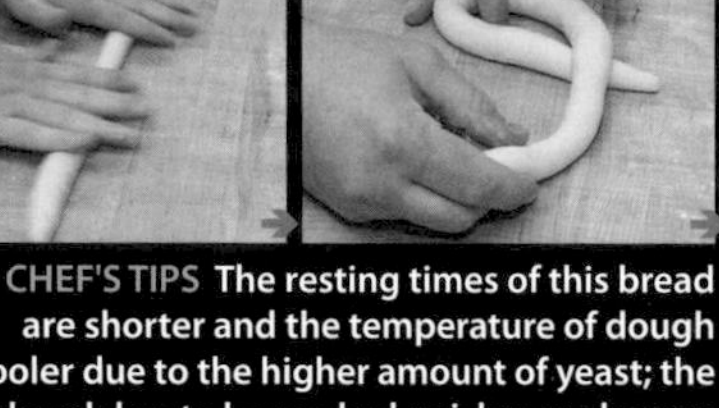

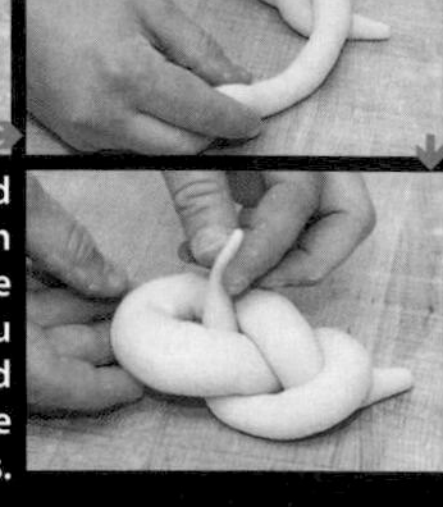

CHEF'S TIPS **The resting times of this bread are shorter and the temperature of dough cooler due to the higher amount of yeast; the dough has to be worked quicker or else you will end up with large bubbles in the finished bread. Eventually finish it with some sesame seeds.**

麵包師建議 **由於含有高量酵母，這麵包的鬆弛時間較其他麵糰短，溫度也較低，所以製作麵糰必須運作很快，否則麵包最終會出現大氣泡。完成製作這麵包後，可撒上少量芝麻作裝飾。**

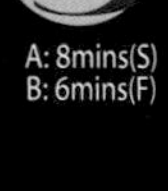
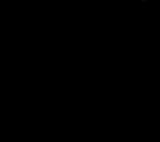
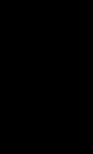

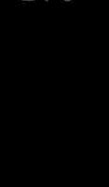

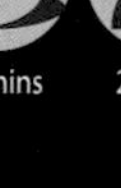
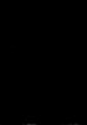

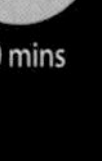
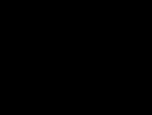
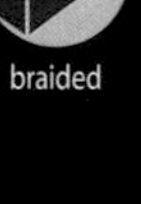
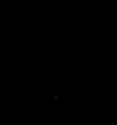
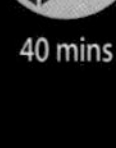

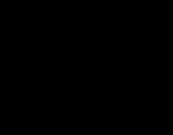

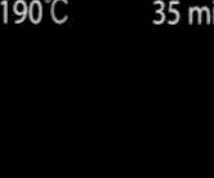

60°C | A: 8mins(S) B: 6mins(F) | 24°C | 30 mins | 250 gm | 10 mins | braided | 40 mins | 200°C | 190°C | 35 mins

	INGREDIENTS	RATIO
440 gm	Flour type 45	100%
200 ml	Milk	45.45%
10 gm	Dry yeast	2.27%
40 gm	White sugar	9.09%
6 gm	Salt	1.36%
40 gm	Eggs	9.09%
60 gm	Butter	13.64%
SQ	Vanilla essential oil	
	Dusting flour	
	Egg wash	
75gm	Candied Ginger	

	材料	比例
440 克	低筋麵粉	100%
200 毫升	鮮奶	45.45%
10 克	乾酵母	2.27%
40 克	細砂糖	9.09%
6 克	鹽	1.36%
40 克	雞蛋	9.09%
60 克	奶油	13.64%
	香草精適量	
	麵粉適量（撒面）	
	蛋液適量（刷面）	
75 克	蜜餞薑粒	

METHOD

1 Knead the dough and add the salt 3 minutes prior kneading ends.

2 After the resting time, shape the dough in braid shape.

3 Place on baking tray with silicon paper.

4 Once fully proofed, brush the loaves with egg wash.

5 Bake until golden brown with steam; open exhaust once coloring starts.

製 法

1 搓揉麵糰，在搓揉過程的最後 3 分鐘，加鹽續搓揉至完成。

2 鬆弛麵糰後，整型成辮子形。

3 在烤盤鋪上矽膠紙。

4 當麵糰發酵後，刷上蛋液。

5 放入蒸氣入烤爐，烘焙至金黃色，待麵包開始轉色，立即開啟排氣活門。

粿麥雜糧麵包

Seeded Oval

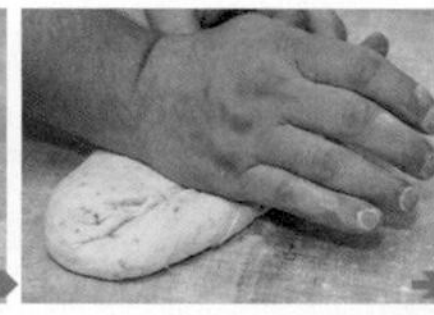

CHEF'S TIPS **Seeds mix can be made of sunflower seeds, sesame, cracked wheat, flaxseed, pumpkin seeds or oat kernel and so on, depending on your taste.**

The fermented dough should be leftover dough from the day before stored in the fridge. If you don't have it on hand, you can produce small dough. To produce the fermented dough, you can use 150 gm of flour type 65 and 100 ml of water with 2 or 3 grams of yeast, knead it and leave it to ferment for 2 hours at room temperature; keep it in the fridge until the next day.

麵包師建議 **混合種籽可由向日葵花種籽、芝麻、壓碎的小麥、亞麻籽、南瓜籽或燕麥粒等組成，按個人喜好便可。**

已發酵麵種應該預早一天準備，貯放在冰箱中備用。如果沒有在手，就可以製造少量麵糰。取用高筋麵粉約 150 克、100 毫升清水、2 或 3 克酵母，搓揉成糰，待室溫下發酵至 2 小時，放進冰箱就可在翌日使用。

A: 10mins(S) B: 8mins(F) | 25℃ | 2 hrs | 350 gm | 20 mins | oval | 1 hr | 225℃ | 200℃ | 45 mins

	INGREDIENTS	RATIO
300 gm	Flour type 65	100%
65 gm	Whole wheat flour	
65 gm	Rye flour	
150 gm	Fermented dough	34.88%
5 gm	Dry yeast	1.16%
75 gm	Seeds mix	17.44%
15 gm	Salt	3.49%
360 ml	Water	83.72%
SQ	Dusting flour	

	材料	比例
300 克	高筋麵粉	100%
65 克	全麥麵粉	
65 克	粿麥麵粉	
150 克	麵種	34.88%
5 克	乾酵母	1.16%
75 克	混合種籽	17.44%
15 克	鹽	3.49%
360 毫升	清水	83.72%
	麵粉適量（撒面）	

METHOD

1. Knead the dough and add the salt 3 minutes before kneading ends.
2. Add the grains in slow speed once the dough is fully kneaded.
3. After the resting time, weigh the dough and pre-shape ovals; rest it on the bench.
4. Shape the ovals and place them on a floured fermenting cloth.
5. Once fully proofed, give one large cut on top of the loaf and bake directly on baking stone.
6. Bake with steam and closed exhaust. Open the exhaust once coloring starts.

製 法

1. 搓揉麵糰，在搓揉過程的最後 3 分鐘，加入鹽續搓揉至完成。
2. 麵糰搓揉完成，以慢速拌入穀類種籽。
3. 鬆弛麵糰後，量重量，造初型呈橄欖狀，置木板上鬆弛。
4. 再整型成橄欖形，放在已撒麵粉的布上發酵。
5. 發酵完成，在麵糰上面切一刀，直接放在烘焙石烘烤。
6. 放蒸氣於烤爐內，關閉排氣活門，然後烘焙，待麵包轉色，立即開啟排氣活門。

全麥愛爾蘭蘇打麵包

Whole Wheat Irish Soda Bread

CHEF'S TIPS The making of Irish soda bread is quick, as a matter of fact; it falls in the category of quick breads; meaning that no fermentation process is involved since the leavening method is chemical with the baking soda. I enjoy this bread with salty butter and the great Irish farm cheeses. During my time working in Ireland, I was positively submerged by the variety and quality of Irish cheeses.

麵包師建議 製造愛爾蘭蘇打麵包非常快捷簡單。事實上，它屬於速成麵包的類別，因發酵方法由蘇打粉的化學反應而成，並沒有經過正確的發酵程序。我享用這包會伴以鹹味奶油和非常美味的愛爾蘭農場起司。當我在愛爾蘭時，當地的優質起司種類繁多。

62℃

A: 10mins(S)
B: 2mins(F)

24℃

none

450 gm

none

round

30 mins

215℃

195℃

45 mins

	INGREDIENTS	RATIO
200 gm	Flour type 55	100%
300 gm	Whole wheat flour	
12 gm	Baking soda	2.4%
10 gm	Salt	2%
300 ml	Fresh milk	60%
150 gm	Plain yogurt	30%
	Dusting flour	

	材料	比例
200 克	中筋麵粉	100%
300 克	全麥麵粉	
12 克	蘇打粉	2.4%
10 克	鹽	2%
300 毫升	鮮奶	60%
150 克	純味優格	30%
	麵粉適量（撒面）	

METHOD

1. Knead the dough lightly, not as much as regular dough, but until 70% of the kneading.
2. Shape the boule and place them in fermenting wooden basket to give them shape and design of flour.
3. Let it rest for 30 minutes.
4. Flip it out of the wooden basket and score a cross on the bread.
5. Bake without steam and closed exhaust. Open the exhaust once coloring starts.

製 法

1. 把麵糰輕輕搓揉，無需按照正常工序，搓揉時間約 70% 便可。
2. 把圓形麵糰，放在發酵用木質的籃內，此法可令麵糰印出花紋。
3. 置一旁，把麵糰鬆弛 30 分鐘。
4. 把發酵木籃翻轉，倒出麵糰，然後在麵糰上刻“十”字紋。
5. 不用放蒸氣於烤爐內，關閉排氣活門，然後烘焙，待麵包轉色，開啟排氣活門。

義大利拖鞋麵包
Classic Ciabatta

64℃ | A: 10mins(S) B: 8mins(F) | 26℃ | 24 hrs at 5℃ | 350 gm | 10 mins | rectangle loaf | 1 hr | 235℃ | 215℃ | 35 mins

	INGREDIENTS	RATIO
75 gm	Flour type 45	100%
300 gm	Flour type 65	
350 ml	Water	93.33%
5 gm	Dry yeast	1.33%
6 gm	Salt	1.6%
20 ml	Olive oil	5.33%
	Dusting flour	

	材料	比例
75 克	低筋麵粉	100%
300 克	高筋麵粉	
350 毫升	清水	93.33%
5 克	乾酵母	1.33%
6 克	鹽	1.6%
20 毫升	橄欖油	5.33%
	麵粉適量（撒面）	

METHOD

1. Mix and knead the dough; store it in a covered plastic container in the fridge.
2. After 12 hours, give the dough a fold. Keep it in the fridge for the remaining 12 hours.
3. Note that the dough will be very soft – use enough flour to dust tables and containers.
4. Flatten the dough to 3 cm thick; allow resting 10 minutes and cut rectangles of 350 gm.
5. Place on baking tray or fermenting cloth. Again, make sure there is enough flour dusted.
6. Once proofed, bake with steam and closed exhaust. Open the exhaust once coloring starts.

製 法

1. 混合材料和搓揉成糰，放入連蓋的塑膠盒內，置冰箱貯藏。
2. 待 12 小時後，取出麵糰摺覆，續保存在冰箱 12 小時。
3. 注意麵糰將會非常柔軟，所以要使用足夠麵粉撒在枱上和塑膠盒內。
4. 碾平麵糰約 3 厘米厚，鬆弛 10 分鐘，分切成 350 克重的長方形。
5. 放入烤盤或發酵布，再次確定那裏有足夠麵粉。
6. 當發酵完成，放蒸氣入烤爐，關閉排氣活門，然後烘焙，待麵包轉色，立即開啟排氣活門。

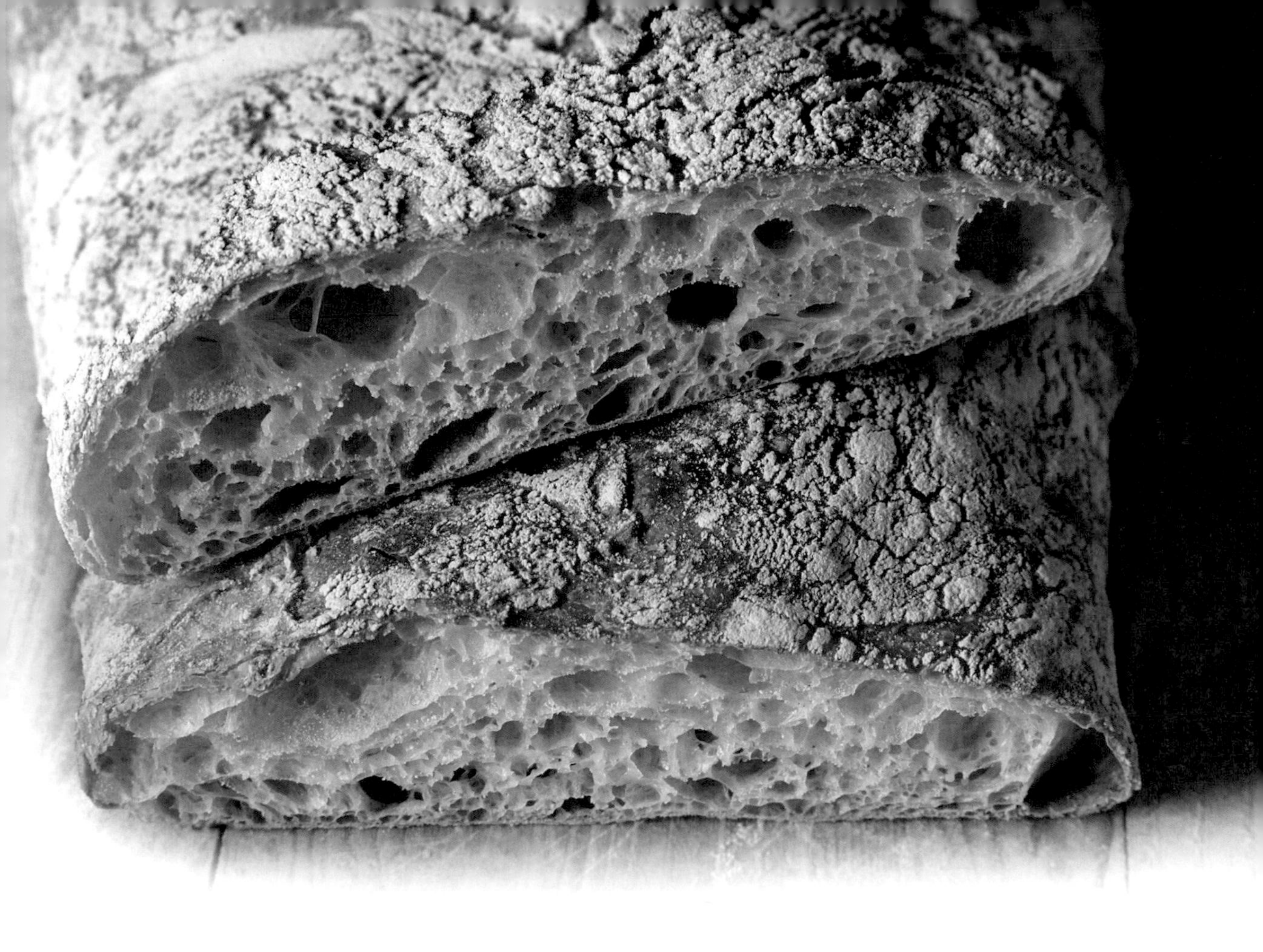

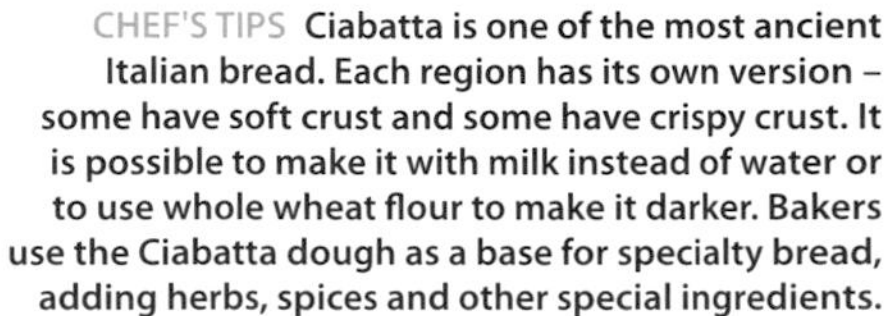
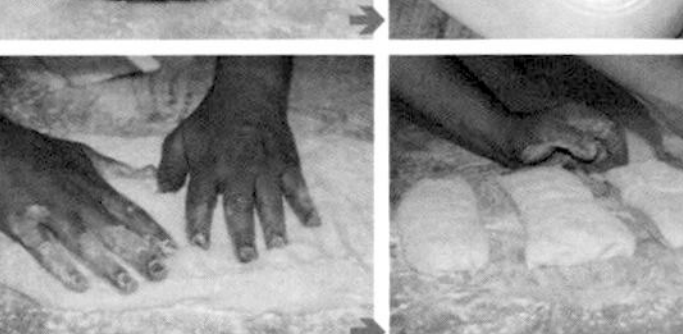

CHEF'S TIPS **Ciabatta is one of the most ancient Italian bread. Each region has its own version – some have soft crust and some have crispy crust. It is possible to make it with milk instead of water or to use whole wheat flour to make it darker. Bakers use the Ciabatta dough as a base for specialty bread, adding herbs, spices and other special ingredients.**

Ciabatta is great for sandwiches or to accompany a nice piece of Parmesan cheese with some virgin olive oil for example. Toasted, the Ciabatta sandwich is known as panino. The key feature of the Ciabatta is its large and fine texture which makes it lighter to eat.

麵包師建議 義大利軟麵包是最古老的義式麵包。每個省份有其獨特款式 —— 有些麵包外層很柔軟；有些含脆香碎屑。可用牛奶替代清水，或可用全麥麵粉，令其變深色。麵包師亦會以香草、香料和其他特別材料作為做特色麵包的基礎。

品嚐義大利軟麵包，其中一種最佳食法是做三明治或可配一片優質帕爾馬起司和初榨純橄欖油。把義大利軟麵包烤成吐司稱為帕尼洛三明治。軟麵包的主要特色是份量大而質感幼細，入口輕軟。

64℃ | A: 10mins(S) B: 8mins(F) | 25℃ | 45 mins | variable | 15 mins | cut letter | 45 mins | 210℃ | 200℃ | 30 mins

	INGREDIENTS	RATIO
500 gm	Flour type 45	100%
140 gm	Butter	28%
50 gm	Sugar	10%
10 gm	Salt	2%
15 gm	Dry yeast	3%
280 ml	Fresh milk	56%
1 pc	Lemon zest	
130 gm	Raisins	26%
SQ	Egg wash	
	Dusting flour	

	材料	比例
500 克	低筋麵粉	100%
140 克	奶油	28%
50 克	細砂糖	10%
10 克	鹽	2%
15 克	乾酵母	3%
280 毫升	鮮奶	56%
1 個	檸檬皮	
130 克	葡萄乾	26%
	蛋液適量（刷面）	
	麵粉適量（撒面）	
	巧克力粒適量	

METHOD

1. Knead the dough and add the salt 3 minutes before kneading ends.
2. Add the raisins and lemon zest in slow speed once the kneading is done.
3. Once rested, fold the dough once and flatten it to about 2 cm in thickness.
4. Place on a floured tray, cover with plastic film and allow to rest in the fridge for 1 hour.
5. Cut the shapes using large letter cutters and place them on a tray with baking paper.
6. Allow proofing and brush egg wash once. Let it dry a little and apply egg wash once more.
7. Bake golden brown with steam and closed exhaust. Open the exhaust once coloring starts.

製 法

1. 搓揉麵糰，在搓揉過程的最後 3 分鐘，加入鹽續搓揉至完成。
2. 麵糰搓揉完成，以慢速拌入葡萄乾和檸檬皮。
3. 鬆弛麵糰後，把麵糰碾平一次，做出約 2 厘米厚。
4. 放入已撒麵粉的烤盤，蓋上保鮮膜，貯放於冰箱 1 小時。
5. 用大字模模鋑出字形，放在已鋪烘焙紙的烤盤上。
6. 讓其發酵，刷上一次蛋液，待乾，再次刷上蛋液。
7. 放入蒸氣於烤爐，關閉排氣活門，然後烘焙至金黃色，待麵包開始轉色，立即開啟排氣活門。

造型麵包
CRUMB

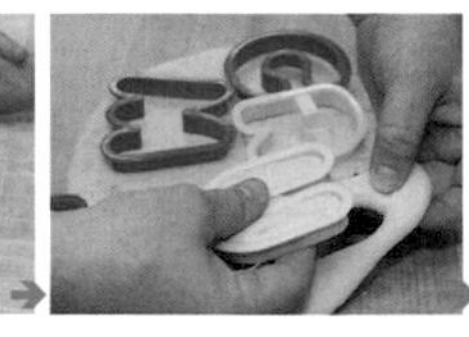
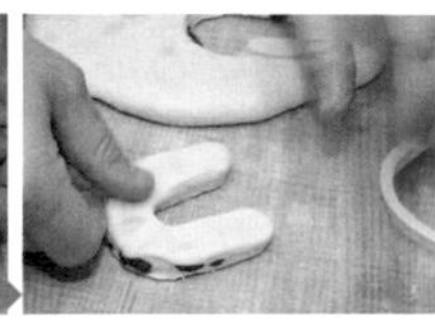

CHEF'S TIPS You can roll the dough thinner and cut smaller letters for example for mini sandwiches or to float on a soup. Great bread for kid' s birthday parties!

Allowing the dough to rest in the fridge will give you sharp shapes and sharp cuts as the dough will not be deformed like it would be if cut right after the dough is flatten. Twice the egg wash is giving you a shiny look, but be careful at baking, it colors faster than usual.

麵包師建議 你可以把麵糰擀平一點，然後鈒成小字母，可做出迷你三明治或讓它浮在湯上。必定能為兒童生日派對帶來驚喜。

讓麵糰置冰箱進行鬆弛，可在切割時令其造型輪廓鮮明，不會變形。刷上蛋液兩次，可為麵包帶光澤，但烘焙時要小心，因上顏色的速度會比平常快。

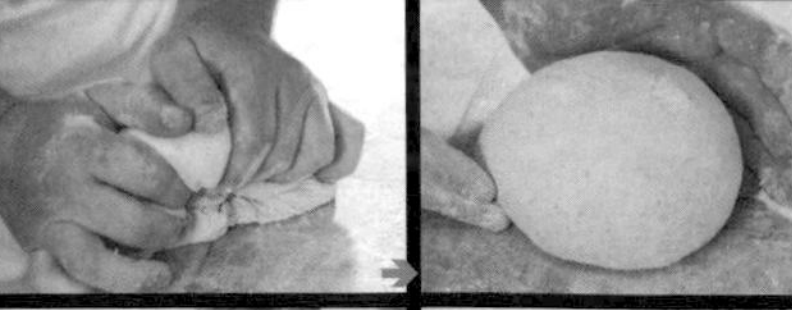

CHEF'S TIPS The country bread has a typical taste of wheat and spelt and goes well with cheese and wine. Toasting the country bread brings out the nutty flavor of wheat and makes it particularly pleasant with cheeses. To obtain a stronger wheat flavor, you can substitute some bread flour and replace it with whole wheat flour, but the texture will increase in density. Indeed, with a more coarse flour mix, there will be less small particles of starch susceptible to ferment into a broad network of gluten. If you wish to have the "hat" of the bread sticking to the bread, brush some water before folding the flatten part over the top, else, the top part will rise.

麵包師建議 鄉村麵包有其獨特的小麥味道，斯佩爾特小麥與起司和酒非常搭配。這麵包經烘烤後帶出小麥裏的果仁味道，因而特別與起司特別匹配。如要獲得濃郁的小麥味道，可把部份麵包粉以全麥麵粉取代，但質感密度會相對地增加。確切地，如麵粉份量多些粗糙感，當發酵至麵筋網絡時便容易受到影響。如希望把“帽子”貼在麵包頂部，把麵糰平坦的部份覆疊在頂部前，先掃上清水，否則麵包部便會在烘烤後上升。

65℃

A: 10mins(S)
B: 6mins(F)

25℃

2 hr

450 gm

15 mins

round

1.5 hrs

235℃

215℃

45 mins

	INGREDIENTS	RATIO
300 gm	Flour type 65	100%
70 gm	Whole spelt flour	
70 gm	Rye flour type 770	
5 gm	Dry yeast	1.14%
15 gm	Wheat germs	3.41%
10 gm	Salt	2.27%
350 ml	Water	79.55%
SQ	Dusting flour	

	材料	比例
300 克	高筋麵粉	100%
70 克	全斯佩爾特小麥麵粉	
70 克	770 號粿麥麵粉	
5 克	乾酵母	1.14%
15 克	小麥胚芽	3.41%
10 克	鹽	2.27%
350 毫升	清水	79.55%
	麵粉適量（撒面）	

METHOD

1 Knead the dough and add the salt 3 minutes before kneading ends.
2 After the resting time, pre-shape in round loaf and allow resting.
3 Shape the loaves and roll thinly one side to fold it over the top. Roll slightly in bread flour.
4 Place on the fermenting cloth up-side-down.
5 Once fully proofed, turn the loaves over and bake directly on baking stone.
6 Bake with steam and closed exhaust. Open the exhaust once coloring starts.

製 法

1 搓揉麵糰，在搓揉過程的最後 3 分鐘，加入鹽續搓揉至完成。
2 麵糰搓揉完成，造初型，呈圓形狀，鬆弛。
3 再修整造型，把一面碾薄，覆上頂面，置麵包粉在工作枱上，輕輕繼續擀平麵糰。
4 反轉麵糰置放在發酵布上。
5 完成發酵後，翻轉麵包，直接放在烘焙石上烘烤。
6 放入蒸氣於爐內，關閉排氣活門，然後烘焙，待麵包開始轉色，立即開啟排氣活門。

斯佩爾特小麥農夫麵包
Spelt Farmer Bread

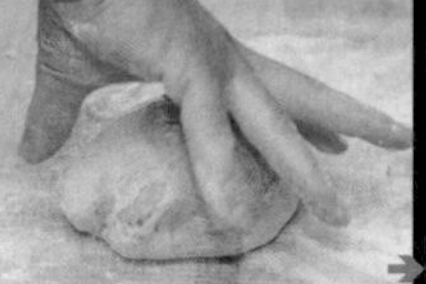

CHEF'S TIPS The acidic and subtle blossom fragrance of Roselle fits very well the earthy flavors of rye. It brings a very pleasant uncommon flavor and new twist to rye bread.

The fermented dough should be leftover dough from the day before stored in the fridge. If you don' t have it on hand, you can produce small dough. To produce the fermented dough, you can use 150 gm of flour type 65 and 100 ml of water with 2 or 3 grams of yeast, knead it and leave it to ferment for 2 hours at room temperature; keep it in the fridge until the next day.

麵包師建議 帶酸和似有若無的洛神花花香，與含土壤味道的裸麥相當搭配。它能帶出不平凡又令人愉快的味道，是粿麥麵包的新搭擋。

麵種應該預早一天準備，貯放在冰箱中備用。如果沒有預先準備，就可以製造少量麵糰。採用高筋麵粉約 150 克、100 毫升清水、2 或 3 克酵母，搓揉成糰，待室溫下發酵 2 小時，放進冰箱就可在翌日使用。

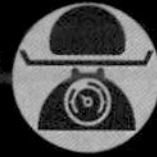
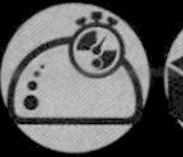

65°C | A: 10mins(S) B: 6mins(F) | 25°C | 2 hr | 400 gm | 15 mins | round | 1.5 hrs | 235°C | 215°C | 40 mins

	INGREDIENTS	RATIO
300 gm	Flour type 65	100%
150 gm	Rye flour	
2 gm	Dry yeast	0.44%
100 gm	Fermented dough	22.22%
12 gm	Sea salt	2.67%
100 gm	Roselle jam	22.22%
375 ml	Water	83.33%
SQ	Dusting flour	

	材料	比例
300 克	高筋麵粉	100%
150 克	粿麥麵粉	
2 克	乾酵母	0.44%
100 克	麵種	22.22%
12 克	海鹽	2.67%
100 克	洛神花果醬	22.22%
375 毫升	清水	83.33%
	麵粉適量（撒面）	

METHOD

1 Knead the dough and add the salt 3 minutes before kneading ends.
2 Add the Roselle jam after the kneading is done, in slow speed.
3 After the resting time, pre shape in round loaf and let it rest on the bench.
4 Shape the loaves and place them up-side-down in wooden proofing basket.
5 Once fully proofed, turn the loaves over and bake directly on baking stone.
6 Bake with steam and closed exhaust. Open the exhaust once coloring starts.

製 法

1 搓揉麵糰，在搓揉過程的最後 3 分鐘，加入鹽續搓揉至完成。
2 麵糰搓揉完成，以慢速拌入洛神花果醬。
3 鬆弛麵糰，做成圓麵糰，放在木板上鬆弛。
4 再造型，把麵糰翻轉在木質發酵籃。
5 讓其完全發酵，翻轉麵包，直接放在烘焙石上烘烤。
6 放入蒸氣於烤爐中，關閉排氣活門，然後烘焙，待麵包開始轉色，立即開啟排氣活門。

洛神花粿麥麵包
Light Roselle Rye Boule

孜然波爾多皇冠麵包

Caraway Bordelaise Crown

CHEF'S TIPS Classic caraway and rye bread, the shape of the Bordelaise crown is nice and attractive. Try not to pull the triangle too hard under each roll, as they will end up deforming during the baking. Instead, cut a larger disc of dough to allow plenty of coverage.

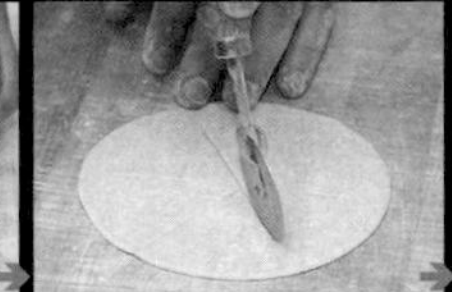

麵包師建議 用傳統孜然和粿麥麵包，做成皇冠狀是非常優美和吸引人。不要嘗試過度用力拉扯至小卷下的三角麵皮，因這樣做會在烘焙時令它變形。取而代之，可切割大塊的圓麵糰，讓其有多些覆蓋面。

64°C | A: 10mins(S) B: 8mins(F) | 25°C | 1 hr | 350 gm total | 30 mins | rolls in crown | 1 hr | 235°C | 215°C | 35 mins

	INGREDIENTS	RATIO
150 gm	Flour type 65	100%
150 gm	Rye flour type 770	100%
240 ml	Water	80%
4 gm	Dry yeast	1.33%
10 gm	Liquid malt	3.33%
2 gm	Caraway seeds	0.67%
8 gm	Salt	2.67%
SQ	Dusting flour	

	材料	比例
150 克	高筋麵粉	100%
150 克	770 號粿麥麵粉	100%
240 毫升	清水	80%
4 克	乾酵母	1.33%
10 克	麥芽糖液	3.33%
2 克	孜然（葛縷子）	0.67%
8 克	鹽	2.67%
	麵粉適量（撒面）	

METHOD

1. Knead the dough and add the salt 3 minutes before kneading ends.
2. After the bulk time, divide the dough and pre-shape the rolls. Keep a part of dough in the fridge for half an hour to roll the central part.
3. Prepare the proofing basket with an up-side-down bowl in the middle, covered with a cloth.
4. Roll the chilled dough in a disc that covers the bowl surface; before leaning the dough in the basket, cut the triangles on the middle of the disc.
5. Lean your disc on the bowl and shape the rolls. Place them up-side-down on the disc of dough.
6. Brush a little water on the base of the rolls and fold the tip of each triangle under the rolls.
7. Once proofed, flip the basket over and bake directly on baking stone.
8. Bake with steam and closed exhaust. Open the exhaust once coloring starts.

製 法

1. 搓揉麵糰，在搓揉過程的最後3分鐘，加入鹽續搓揉至完成。
2. 麵糰發酵完成，分割麵糰，做成小卷。可預留部份麵糰，存放在冰箱半小時，取出後捲起用作麵包中央。
3. 準備一個發酵用木質籃，將一個碗倒扣於木籃中央位置，蓋上一發酵布。
4. 取出冷凍麵糰，碾平成圓片狀，蓋在碗的表面上。斜放在發酵籃的麵糰前，切成三角，再放在圓麵皮的中央。
5. 放完麵皮，搖動發酵籃，
6. 在麵包底部刷點清水，覆摺三角的尖端，放在底部。
7. 待其完全發酵，把木質籃取出，麵糰翻轉，直接放烘焙石上烘烤。
8. 放入蒸氣，關閉排氣活門，然後烘焙，待麵包開始轉色，立即開啟排氣活門。

 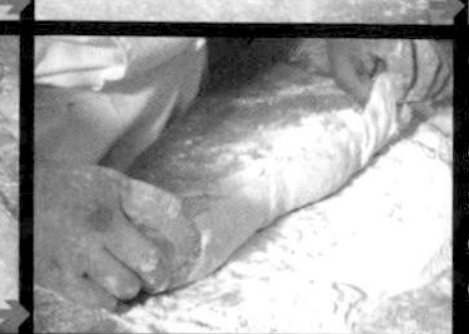

CHEF'S TIPS The ffermentation process technique of more than 24 hours allows this very liquid dough' s gluten to coagulate over time. The dough will still be soft, but when baking it will create a pleasant texture, crumb and crust. This bread has a fairly long time of conservation.

At the early stage of development of this now popular bread in Europe, my good friend Fabrice and our team worked endless hours to develop this version of the loaf into a more matured dough method, giving you this fantastic rustic bread.

麵包師建議 食譜發酵過程用上 24 小時有別於其他方法，這技術可讓麵糰和麵筋非常流質狀，隨時間結在一起。麵糰將仍然柔軟，但烘烤無論麵包外層或內層，均可助製作出很出色的質感、麵包屑和脆碎屑。這種麵包的保存期可較長。

這麵包現已普及於歐洲，但發展初期，我和好友 —— 費比斯和工作團隊，花上無數小時研發出一個較長時間而作做出的麵糰方法，繼續發展出這新款的品種，給予人驚喜的農村麵包。

   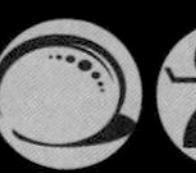     

54°C | A: 8mins(S) B: 8mins(F) | 22°C | 1st 24 hrs at 5°C 2nd 3 hrs at 25°C | 380 gm | 15 hrs | twisted baguette | 40 mins | 245°C | 220°C | 35 mins

	INGREDIENTS	RATIO
600 gm	Flour type 65	100%
25 gm	Whole wheat flour	
35 gm	Rye flour	
100 gm	Fermented dough	15.15%
3 gm	Dry yeast	0.45%
15 gm	Wheat germs	2.27%
15 gm	Salt	2.27%
550 ml	Water	83.33%
SQ	Dusting flour	

	材料	比例
600 克	高筋麵粉	100%
25 克	全麥麵粉	
35 克	粿麥麵粉	
100 克	發酵麵糰	15.15%
3 克	乾酵母	0.45%
15 克	小麥胚芽	2.27%
15 克	鹽	2.27%
550 毫升	清水	83.33%
	麵粉適量（撒面）	

METHOD

1. Knead the dough and add the salt 3 minutes before kneading ends. The dough will be very soft with a liquid texture.
2. Keep the dough in large plastic containers; the dough will triple its size in the next 24 hours.
3. Keep the dough in bulk at 25 ° C for 3 hours; place the dough on the heavily floured table.
4. Carefully, weigh the dough into long shape pieces and twist them in bread flour.
5. Place on the fermenting cloth for 40 minutes.
6. Make sure the breads don' t stick to the cloth and bake directly on the stone.
7. Bake with steam and closed exhaust. Open the exhaust once coloring starts.

製 法

1. 搓揉麵糰，在搓揉過程的最後 3 分鐘，加入鹽續搓揉至完成。麵糰將非常柔軟，帶著流質的感覺。
2. 把麵糰保存在塑膠盒內，24 小時，它會變成大 3 倍。
3. 繼續待在溫度 25°C的地方發酵 3 小時，然後放在撒滿大量麵粉的工作枱上。
4. 小心翼翼地把量麵糰量重，再搓成長條形，在麵包粉上扭扣在一起。
5. 放在發酵布待 40 分鐘。
6. 確保麵包沒有黏貼在布上，然後直接放在烘焙石上烘烤。
7. 放入蒸氣於烤爐內，關閉排氣活門，然後烘焙，待麵包開始轉色，立即開啟排氣活門。

鄉村麵包棒
Fabrice Country Stick

辣根亞麻籽粗麥方塊麵包

Horseradish Flaxseed Semolina Squares

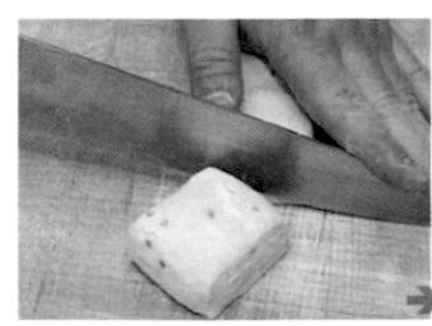

CHEF'S TIPS **The blend of horseradish and flaxseed into semolina bread is unique and a great match to your dining table. Think about serving it with any roasted meat and other hearty meals.**

For the horseradish: Find fresh horseradish roots, peel them and cut them into dice. In a food processor, add a little ice water, salt and a table spoon of vinegar. Blend it until it has the texture of a horseradish cream.

麵包師建議 **融合了辣根和亞麻籽於粗麥粉麵包中，它的味道獨特兼是用餐時的最佳拍擋。嘗試與烤肉享用或配其他愛心美食，均有意想不到的效果。**

關於辣根的製法，可用新鮮辣根的莖部，去皮和切成小丁粒，然後放進搓揉機，加少許冰水、鹽和 1 茶匙醋，搓揉成幼滑質感。

60℃ · A: 8mins(S) B: 6mins(F) · 25℃ · 1 hr · 50 gm · 15 mins · square · 30 mins · 215℃ · 200℃ · 25 mins

	INGREDIENTS	RATIO
500 gm	Fine semolina	100%
10 gm	Dry yeast	2%
50 gm	Butter	10%
12 gm	Salt	2.4%
300 ml	Water	60%
50 gm	Horseradish	10%
10 gm	Flaxseed	2%
SQ	Semolina for dusting	

	材料	比例
500 克	優質粗麥麵粉	100%
10 克	乾酵母	2%
50 克	奶油	10%
12 克	鹽	2.4%
300 毫升	清水	60%
50 克	辣根	10%
10 克	亞麻籽	2%
	粗麥麵粉適量（撒面）	

METHOD

1 Knead the dough and add the salt 3 minutes before kneading ends.

2 Add the horseradish and flaxseed after the kneading process is done, in slow speed.

3 After the bulk time, flatten the dough to about 2 cm thick and let it chill in the fridge for about 1 hour, covered with plastic.

4 Cut the dough slab into squares of about 5 cm and place on a baking tray.

5 Once proofed, bake with steam and closed exhaust. Open the exhaust once coloring starts.

製 法

1 搓揉麵糰，在搓揉過程的最後 3 分鐘，加鹽續搓揉至完成。

2 麵糰搓揉完成後，以慢速拌入辣根和亞麻籽。

3 完成初步發酵麵糰，擀平麵糰約 2 厘米厚，蓋上保鮮膜，放冰箱冷凍 1 小時。

4 把厚身的麵糰再切成約 5 厘米的方塊，置於烤盤上。

5 一旦發酵完成，放入蒸氣，關閉排氣活門，然後烘烤，待麵糰轉色，立即開啟排氣活門。

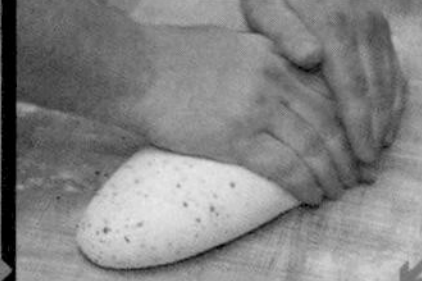

CHEF'S TIPS **You can shape the loaf either in one piece or you can divide it in 2 or 3 pieces. Shape round loaves and place them next to each other in the toast mould.**

The addition of seaweed and lemon gives this loaf a fine finish, giving you new ground for creative tartines and sandwiches.

麵包師建議 **你可以把麵包造型狀條或分成 2 至 3 細小條。甚至造成圓形，只要邊貼邊放在吐司模內便可。**

額外添加海藻和檸檬會在麵包內可令麵包有細緻的效果，希望能給予你製做法式吐司和三明治一些新靈感。

52°C | A: 12mins(S) B: 6mins(F) | 24°C | 1 hr | 550 gm | 30 mins | square toast | 2 hrs | 215°C | 195°C | 45 mins

	INGREDIENTS	RATIO
500 gm	Flour type 45	100%
25 gm	Eggs	5%
7 gm	Dry yeast	1.4%
7 gm	Salt	1.4%
50 gm	Butter	10%
50 gm	Sugar	10%
50 ml	Fresh milk	10%
280 ml	Water	56%
12 gm	Dried seaweed	2.4%
1 pc	Lemon zest	

	材料	比例
500 克	低筋麵粉	100%
25 克	雞蛋	5%
7 克	乾酵母	1.4%
7 克	鹽	1.4%
50 克	奶油	10%
50 克	砂糖	10%
50 毫升	鮮奶	10%
280 毫升	清水	56%
12 克	乾海藻	2.4%
1 個	檸檬皮	

METHOD

1. Knead the dough and add the salt 3 minutes before kneading ends.
2. Add the lemon zest and seaweed at the end of kneading in slow speed.
3. After the resting times, shape the long loaves and place them in the greased toast moulds.
4. Once fully proofed, bake with steam and closed exhaust; open it once the coloring starts.
5. Unmould the baked loaves shortly after the end of baking. Store them on a cooling grid or else condensation will make the base of the bread damp.

製 法

1. 搓揉麵糰，在搓揉過程的最後 3 分鐘，加鹽續搓揉至完成。
2. 麵糰搓揉完成，以慢速拌入檸檬皮和海藻。
3. 完成發酵，搓成長包形，放入已塗油的吐司模。
4. 一旦發酵完成，放入蒸氣於烤爐，關閉排氣活門，然後烘烤，待麵糰轉色，立即開啟排氣活門。
5. 烘烤完成後在短時間內脫模，放在鏤空的不銹鋼網架上貯放，否則吐司底部會因冷凝而變濕潤。

海藻檸檬絲吐司

Seaweed Lemon Pain de Mie

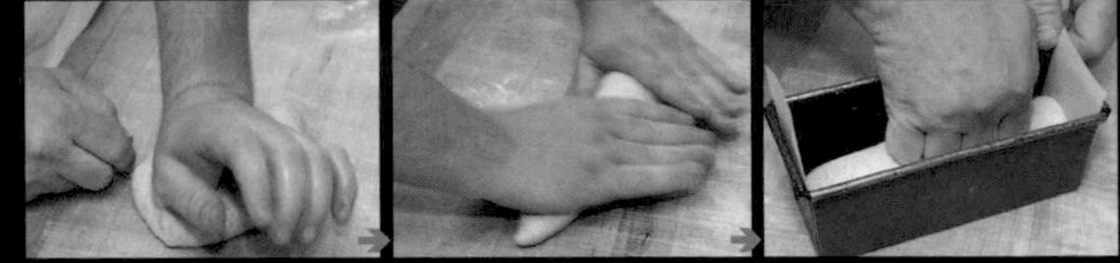

CHEF'S TIPS Adding some wheat bran to the whole wheat loaf will bring a nutty taste when the slice will be toasted for breakfast with jams and butter or with any English breakfast dishes.

To obtain a tighter texture in the loaf, it is possible to shape it divided in 4 parts of dough in round shape placed next to each other in the greased mould. In toast bread, it is important to shape the loaf without air bubble in order to avoid any large holes after baking.

麵包師建議 添加麥麩於全麥麵包可令切成薄片後帶出果仁味道，和更有口感，烘烤後加上果醬、奶油和任何英式早點搭配，便可作一頓豐富的早餐。

想麵包的質地緊密，可把麵糰分成 4 份，造成圓型，邊貼邊放於已塗油的模內。麵包烘烤後，最重要是確保麵糰造型時沒有氣泡，重要避免烘焙後出現大氣泡。

 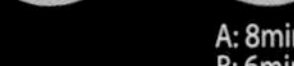    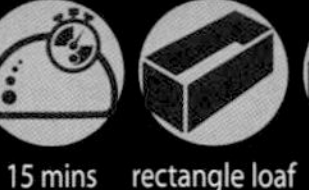     

54℃ | A: 8mins(S) B: 6mins(F) | 25℃ | 1 hr | 450 gm | 15 mins | rectangle loaf | 2 hrs | 215℃ | 190℃ | 40mins

	INGREDIENTS	RATIO
700 gm	Flour type 65	100%
300 gm	Whole wheat flour	
50 gm	Wheat bran	
15 gm	Dry yeast	1.43%
15 gm	Salt	1.43%
100 gm	Butter	9.52%
100 gm	Sugar	9.52%
100 ml	Fresh milk	9.52%
600 ml	Water	57.14%
SQ	Whole wheat flour for dusting	
	Water	

	材料	比例
700 克	高筋麵粉	100%
300 克	全麥麵粉	
50 克	麥麩	
15 克	乾酵母	1.43%
15 克	鹽	1.43%
100 克	奶油	9.52%
100 克	砂糖	9.52%
100 毫升	鮮奶	9.52%
600 毫升	清水	57.14%
	全麥麵粉適量（撒面）	
	清水適量（刷面）	

METHOD

1. Knead the dough and add the salt 3 minutes before kneading ends.
2. After the resting times, shape the loaves and place them in greased loaf moulds.
3. Once fully proofed, brush the top of the loaf with water and sprinkle whole wheat flour.
4. Bake with steam and closed exhaust. Open the exhaust once coloring starts.
5. Unmould the baked loaves shortly after baking and place them on a cooling grid or else condensation will make the base of the bread damp.

製 法

1. 搓揉麵糰，在搓揉過程的最後 3 分鐘，加鹽續搓揉至完成。
2. 完成鬆弛麵糰後，為麵糰造型，放入已塗油的吐司模。
3. 發酵完成後，在麵包面刷點清水，撒上全麥麵粉。
4. 放入蒸氣於烤爐，關閉排氣活門，然後烘烤，待麵糰轉色，立即開啟排氣活門。
5. 烘烤完成後短時間脫模，放在涼凍架上貯放，否則吐司底部會因冷凝而變濕潤。

Glossary of baking terms
烘焙術語彙編

Blade 刀片	Either razor blade, knife or special blade used to score the bread. 任何一款鬚刨剃刀、小刀或特別用於麵包刻紋的刀片。
Blister 水泡	Small bubbles forming on the crust of the bread while baking. 烘焙時，形成於麵包外層的小氣泡。
Body 主體	Used to define the texture of dough, either firm or soft or batard (neither firm nor soft). 用於確定麵糰的質地堅硬、柔軟或軟硬適中。
Bulk proof/ bulk fermentation 初次發酵 / 初步發酵	The first proofing time after the mixing and kneading of dough. 混合和搓揉麵糰後，第一次發酵的時間。
Burned 燒焦 / 過度烘焙	Said of bread that is over baked underneath, with a too high temperature of the stone bed. 麵包在烘焙底層石板的溫度太高，令麵包過度烘焙。
Burnt dough 麵糰燃燒	Over mixed dough, passed the peak of gluten formation. 過度搓揉麵糰，超越麵糰形成筋性的極限。
Cloth 布	Piece of cloth used to ferment the shaped loaves. 一塊用於蓋住已造型麵糰以備發酵的布。
Collapse 倒塌 / 瀉下	Dough which collapses after mixing or shaping due to a lack of strength. 麵糰因混合或造型缺乏韌度而下塌。
Cover 覆蓋 / 蓋面	To cover dough to avoid drying out. 覆蓋麵糰表面，防止風乾。
Dividing 出體（麵糰分割）	To divide dough after the bulk fermentation. 麵糰發酵後分成小糰，行內術語是出體。
Falling 降溫	Said of an oven with temperature decreasing from the start of baking. 泛指烤爐的溫度由開始烘焙時而降低。
Final proof 發酵 / 最後發酵	The last proof after shaping and before baking. 成形後入爐前的最後發酵。
Flat 平坦	Said of a loaf that lack of strength due to too soft dough or weak flour. 泛指麵糰因太軟或麵粉太弱而缺乏力量。
Floured 撒粉	To dust flour on a working space, a fermentation cloth or a dough. 在工作枱、發酵布或麵糰上撒麵粉。
Folding 複摺	To fold dough in order to extract the carbonic cells to give more strength to the dough. 麵糰摺複為了額外的二氧化碳空間，給予麵糰更強韌度。
Kneading 搓揉	To stretch and pull a dough to create the gluten network. 伸展或拉扯麵糰以生產麵筋網絡。

Mixing 混合	To homogenize the ingredients to form dough before the kneading process. 在搓揉過程前，將不同的材料形成一團。
Refresh 恢復 / 回復	To add flour and water, especially to sourdough, to further continue the fermentation. 加麵粉和清水，特別是酸酵麵種，目的是進一步繼續發酵。
Resting 鬆弛 / 鬆身	To allow the divided pieces of dough to rest before shaping. 讓已分體的麵糰在造型前鬆身。
Rise 升起	Said of dough or loaf when it is expanding. 指麵糰或麵包膨脹。
Roll 捲	The action of shaping a round loaf. 造圓形麵糰或麵包的動作。
Scoring 刻紋 / 劃紋	Using different blades to score the breads before baking. 烘焙前用不同刀片在麵包上刻紋。
Shaping 造型	To shape a piece of dough into the required shape while extracting gases. 發酵麵糰之時按照指定形態造型。
Skin 脫皮	A skin formed following the lack of cover on dough. 麵糰沒有蓋面，而形成皮層。
Sole 磚爐	The bottom of the oven usually made of real stone, nowadays replaced by stone composite. 烤爐的底部一般附有真石頭，時至今日，以由合成石頭取代真石頭。
Steam 放入蒸氣	Water injected in the oven to wet the outer layer of the bread in order to allow expansion. 在烤爐噴出清水，令麵糰表面濕潤，進一步膨脹。
Strength 力度 / 韌度	Said of the relative power of dough or loaves. 麵糰或麵包的相關柔韌度。
Tight dough 收緊麵糰	Add flour during the mixing to make the dough firmer. 當混合時加入麵粉，令麵糰堅硬。
Tight shaping 收緊造型	To shape a loaf with more strength relative to the texture of dough and desired end product. 麵糰造型時多加點力度，使最後的製品有緊密質地。
Water Cooler 水冷卻器	Piece of equipment used to cool water where temperature is too high. 一具儀器，用作冷卻過高溫度。
Welding 接合處	The closing point of a loaf that is usually placed underneath the shaped dough. 麵糰的收口 / 接合位，常於已造型麵糰的底部。
Young dough 年輕麵糰	Said of dough that is not rested enough to carry on to the next step. 麵糰鬆弛不足，在下一步驟繼續。

Acknowledgements
鳴謝

My heartfelt thank you and most respected thoughts go to Mark Yeung: Hong Kong's finest baker! Ringo Chan, Herve Fucho, Vincent Thierry, Celene Loo, Xavier Honorin and Saravanan Raman for their great input. And of course, to make everything possible, a big thanks to the unique bakery and pastry team of the Four Seasons Hotel Hong Kong.

But most of all and for the greatest support, all my love goes to Vianna, Clement, Papa, Maman, Sebastien & Laetitia, Lai Ying and New.

我衷心地感謝香港最佳麵包師楊嘉明、陳永雄、 Hervé Fucho，Vincent Thierry，Celene Loo， Xavier Honorin and Saravanan Raman給予我很多寶貴意見。當然， 還要特別多謝四季酒店獨一無二的糕餅房團隊 ，沒有你們的支持，便不能成功。

最重要還是感謝我的太太 Vianna、兒子 Clément、爸爸、媽媽、弟弟 Sébastien、弟婦 Laetitia、岳母麗英和大舅健成，給予無盡的愛和支持。

古風歐陸麵包

La Boulangerie-Baking at home with Grégoire Michaud

作　　者　閔言樂
發 行 人　程安琪
總 策 劃　程顯灝
總 編 輯　潘秉新
執行總編輯　錢嘉琪
文字統籌　燕子老師
攝　　影　幸浩生
封面設計　李傳慧

出 版 者　橘子文化事業有限公司
總 代 理　三友圖書有限公司
地　　址　106 台北市安和路2段213號4樓
電　　話　（02）2377-4155
傳　　真　（02）2377-4355
E-mail　service @sanyau.com.tw
郵政劃撥　05844889 三友圖書有限公司

總 經 銷　大和書報圖書股份有限公司
地　　址　新北市新莊區五工五路2號
電　　話　(02)8990-2588
傳　　真　(02)2299-7900

國家圖書館出版品預行編目資料

古風歐陸麵包 / 閔言樂作. -- 初版. --
臺北市：橘子文化, 2012.09
　面；　公分

ISBN 978-986-6062-24-7(平裝)
1.點心食譜　2.麵包

427.16　　101017516

初　　版　2012年10月
定　　價　新臺幣380元
ISBN　978-986-6062-24-7 (平裝)